Praise for William Pou

The Doomsday Calculation

How an Equation that Predicts the Future Is Transforming Everything We Know About Life and the Universe

"One of the best science writers of our time has taken on one of the most interesting and important subjects of all time—how to predict the future under great uncertainty. What are the odds that a nuclear weapon will detonate by accident? Is there a chance that humanity will destroy itself this century? What is the probability that aliens are out there? Will we colonize the galaxy? Are we living in a computer simulation? Is your marriage or partnership about to come to an end? These seemingly unrelated questions may be answered through one equation and a handful of principles that, when applied, may very well save humanity. A gripping read."

—Michael Shermer, publisher of *Skeptic* magazine and author of *Heavens on Earth*

"Poundstone marshals arguments for and against the plausibility of doomsday predictions with vivid and subtle ease... Demonstrates applications of doomsday-style arguments in finance, quantum physics, the potential threat of artificial intelligence, and the question of whether we might all be living in a computer simulation."

—Steven Poole, *Wall Street Journal*

"An amusing, entertaining effort to answer the unanswerable. Poundstone's examples mix statistics and serious philosophical arguments, and readers who pay close attention will be rewarded."

—*Kirkus Reviews*

The Doomsday Calculation

Also by William Poundstone

Head in the Cloud

Rock Breaks Scissors

Are You Smart Enough to Work at Google?

Big Secrets

The Recursive Universe

Bigger Secrets

Carl Sagan: A Life in the Cosmos

How Would You Move Mount Fuji?

Fortune's Formula

Gaming the Vote

Priceless

The Doomsday Calculation

HOW AN EQUATION THAT PREDICTS THE FUTURE IS TRANSFORMING EVERYTHING WE KNOW ABOUT LIFE AND THE UNIVERSE

WILLIAM POUNDSTONE

Little, Brown Spark
New York Boston London

Little, Brown Spark
Hachette Book Group
1290 Avenue of the Americas, New York, NY 10104
littlebrownspark.com

Originally published in hardcover by Little, Brown Spark, June 2019
First Little, Brown Spark paperback edition, August 2020

Little, Brown Spark is an imprint of Little, Brown and Company, a division of Hachette Book Group, Inc. The Little, Brown Spark name and logo are trademarks of Hachette Book Group, Inc.

ISBN 978-0-316-44070-7 (hardcover) / 978-0-316-42392-2 (international) /
978-0-316-44069-1 (paperback)
LCCN 2018957418

10 9 8 7 6 5 4 3 2 1

LSC-C

Printed in the United States of America

To Arthur Saint-Aubin

Time is a game played beautifully by children.

—*Heraclitus*

He marveled at the fact that cats had two holes cut in their fur at precisely the spot where their eyes were.

—*Georg Christoph Lichtenberg*

Contents

The Doomsday Calculation

Mark Tansey, *Achilles and the Tortoise*, 1986. © Mark Tansey

Diana and Charles

Diana Spencer met Charles, Prince of Wales, at a garden party in 1977. The couple fell in love and, after due diligence by their families, wed at St. Paul's Cathedral in July 1981.

American artist Mark Tansey incorporated Diana in his 1986 painting *Achilles and the Tortoise.* She is shown planting a hemlock, a sapling version of the mature tree behind her. Diana was often photographed planting trees, among them an apple tree she planted in honor of Isaac Newton.

In 1993 Diana came to the attention of American astrophysicist J. Richard Gott III. Gott had devised a mathematical formula for predicting the future. He wanted to test it on a celebrity marriage, and he chose Charles and Diana's because a magazine reported they were the most famous couple of the time. Gott's formula predicted a 90 percent chance that the royal marriage would end in as little as 1.3 more years. At the time, a royal divorce was considered almost unthinkable.

In December 1995 Queen Elizabeth II, incensed by tabloid reports of the couple's extramarital affairs, wrote a letter advising Charles and Diana to divorce. The split was formalized on August 28, 1996. The following year, on August 31, 1997, Diana had a champagne supper in Paris with her new romantic interest, film producer Dodi Fayed. After leaving the restaurant, Diana and Dodi were killed when their alcohol-impaired chauffeur challenged paparazzi to a street race.

Tansey's picture contains at least four other portraits. To the right of Diana is mathematician Mitchell Feigenbaum holding a bottle of champagne, whose bubbles epitomize chaos theory. Feigenbaum, a pioneer of that theory, demonstrated that many phenomena are fundamentally unpredictable. In 1996 he founded Numerix, a firm using Bayesian probability to price financial derivatives for the so-called rocket scientists of Wall Street.

To the right of Mitchell, though easily missed, is the familiar face of Albert Einstein, shown in profile. The speeding rocket and slow-growing hemlock allude to Einstein's thought experiments of racing trains and light beams, used to develop his theory of relativity. Standing in front of Einstein is Benoit Mandelbrot, the IBM mathematician who described the concept of fractals. The hemlock tree and rocket blast are fractals, complex shapes in which each part resembles the whole.

Zeno of Elea, a Greek philosopher whose features are known from ancient busts, dangles a cigarette. Zeno propounded the paradox of Achilles and the Tortoise. Swift Achilles challenges the Tortoise to a footrace. The Tortoise demands a head start. Whenever Achilles catches up to where the Tortoise was, he still has a little farther to go. Thus, Zeno argued, Achilles can never overtake the Tortoise. For Zeno's followers, the paradox was proof that something is deeply wrong about our understanding of space, time, and reality.

This book tells the story of another mind-boggling idea, the doomsday argument. As advanced by Gott and other scholars, it is a mathematical scheme to predict how long the human race will survive. The idea seems incredible to almost everyone at first encounter, but as we will see, it is not easily dismissed. In the following chapters I will present the cases for and against this provocative idea and attempt to evaluate them. I will show how the type of reasoning used in the doomsday argument has many potential applications. The argument has caused bright people to reflect on our fragile existence, our hopes, and our obligation to future generations—and to reexamine the nature of evidence and the place of humans in the universe.

Part I

Consider the Lemming

The end is near. Or not. The following chapters explore the doomsday argument, a simple line of reasoning that leads headlong to the conclusion that humanity does not have much time left. We meet the doomsayers and their critics and encounter such topics as the runs of Broadway plays, the populations of lemmings, and the riddle of Sleeping Beauty. We find that at least some doomsday calculations deserve to be taken seriously, and we assess our prospects.

How to Predict Everything

Six-year-old Helen Gregg, her nine-year-old sister, Frances, and their nine-year-old cousin, Ella Davies, never saw the atomic bomb that hit their playhouse. They were about six hundred feet away, in the South Carolina woods, on that bright spring day of March 11, 1958. The bomb was egg-shaped with stabilizing fins, a near-twin of the "Fat Man" bomb that struck Nagasaki. It annihilated the playhouse that Helen and Frances's father had built for the girls, leaving a crater seventy-five feet across and thirty feet deep.

All the tons of earth thrown up in the air came back down in a hellish rain. It was that that injured the three girls, parents Walter and Effie Gregg, and their son Walter Jr. There were no deaths aside from a few chickens. The Greggs lived in a town called Mars Bluff. Today, sixty summers later, the crater is still visible.

Albert Madansky was a young statistics PhD from the University of Chicago, recruited by the RAND Corporation, a Santa Monica think tank contracting to the Pentagon. RAND wanted Madansky to tackle a problem that was easy to state but difficult to answer: What is the probability of a nuclear weapon detonating by accident?

The Mars Bluff incident, occurring the year after Madansky

began work at RAND, was a prime topic of discussion. Madansky learned what the public had not. A B-47 Stratojet had left Hunter Air Force Base, Georgia, as part of a drill in handling atomic weapons. Early in the flight a red warning light came on in the cockpit, indicating that the bomb wasn't properly secured.

Copilot Bruce Kulka banged the warning light with the butt of his service revolver. The light went off. Later it came back on. Kulka went to the bomb bay to fix the problem. He reached around the bomb to engage a lock, hitting the wrong button. The weapon came loose, crashing through the bomb bay doors and plummeting fifteen thousand feet.

A fission bomb contains chemical explosives, TNT in this case, surrounding a core of uranium or plutonium. Unspeakable tragedy was avoided only because the bomb was unarmed, without any fissile material. The ground impact detonated the TNT, however, creating a massive conventional explosion.

Accidents like Mars Bluff had been happening for some time. Madansky was allowed to see a top secret list of sixteen "dramatic incidents" that had occurred between 1950 and 1958.

RAND's people worried about other scenarios. What if a bomb was lost and a civilian found it? What if an angry or unstable officer launched an atomic bomb without authorization? There were no statistics on such events because they had never happened.

In conventional statistical thinking, you can't assign a probability to something that has never happened. Whereof one has no data, one must remain silent.... But Madansky had studied statistics at Chicago with Leonard "Jimmie" Savage. Savage had been born with the name Ogashevitz, though it was generally agreed that Savage fit him better. He was brutally critical of anyone he judged less brilliant than himself, a group that seemed to cover just about everyone in the fields of mathematics and economics. Savage was a contrarian by nature. One of his most contrary pet ideas was Bayes's theorem—an obscure formula, named for an obscure minister of eighteenth-century England. Madansky was able to see that Bayes's theorem offered exactly what RAND needed: a way to assign a probability to doomsday.

RAND's 1958 report (authored by Madansky and colleagues Fred Charles Iklé and Gerald J. Aronson, and declassified in 2000) noted that the US atomic arsenal was growing rapidly, multiplying the opportunities for an accident. At the height of the Cold War, the Strategic Air Command intended to keep about 270 B-52 bombers in the air at all times, ready to launch a nuclear attack on word from the president.

"A probability that is very small for a single operation, say one in a million, can become significant if this operation will occur 10,000 times in the next five years," the RAND report warned. With more bombs being transported more miles, the authors computed that a major catastrophe was near-inevitable in just a few years.

The report sketched countermeasures, ranging from the mundane to the bizarre. It proposed electrifying the bomb's arming switches, so that anyone touching them would get a mild shock, lessening the chance of accidentally hitting the wrong button. As to the *Dr. Strangelove* scenario of a deranged individual starting World War III, the report argued for psychological screening of all who worked with the bombs. The most practical ideas were to put combination locks on bombs and to arrange that two individuals must act simultaneously to arm a bomb.

The RAND group was reporting to General Curtis LeMay, a no-nonsense war hero who fretted about American leadership being too politically correct to use its nuclear weapons. To Madansky's relief, LeMay immediately grasped the seriousness of the problem. The general ordered the combination locks and the two-person system.

In folk wisdom, lightning never strikes the same place twice. Yet on January 24, 1961, the Carolina low country had another nuclear close call. One of LeMay's B-52s developed a fuel leak and began to break up in midair near Goldsboro, North Carolina. As the tail sheared off, two bombs slid out of the bomb bay and plunged to earth. Three crew members died, and five parachuted to safety.

There wouldn't have been *any* safety had the bombs gone off. This B-52 was carrying hydrogen bombs. Had either of them detonated, the fallout plume would have reached Philadelphia.

One of the bombs was discovered suspended from a tree by its parachute. It had barely kissed the earth. The "arm/safe" switch was still on "safe."

The other bomb's parachute failed to deploy. This bomb broke apart, and the fragments fell into a swampy area with enough water to soften the impact and spare the conventional explosives.

Bomb disposal expert Lieutenant Jack ReVelle was called in to find the pieces. "Until my death," ReVelle said, "I will never forget hearing my sergeant say, 'Lieutenant, we found the arm/safe switch.' And I said, 'Great.' He said, 'Not great. It's on "arm."' "

"You're the Product"

Thomas Bayes, the nonconformist minister of Tunbridge Wells, England, drew his last breath on April 17, 1761. For reasons not clear he left his life's greatest achievement filed away, unpublished and unread. It was another mathematically inclined minister, Richard Price, who found Bayes's manuscript after his death and recognized its importance. Price counted among his acquaintances a notorious group: the American revolutionaries Thomas Paine, Thomas Jefferson, and Benjamin Franklin, as well as Mary Wollstonecraft, the feminist who married an anarchist and gave birth to the author of *Frankenstein*.

Price sent the Royal Society of London "an essay which I have found among the papers of our deceased friend Mr. Bayes, and which, in my opinion, has great merit."

This essay described what we now call Bayes's theorem (or rule or law). It addresses a fundamental question of the Enlightenment worldview: How do we adjust our beliefs to account for new evidence?

To put it in modern terms, you start with a *prior probability* ("prior," for short). This is an estimate of the likelihood of something happening, based on everything already known. This estimate is then adjusted up or down for new data, according to a simple formula.

Price praised Bayes's ingenuity but offered this warning: "Some of the calculations... no one can make without a good deal of labour."

Partly for that reason Bayes's theorem was neglected. Repeated calculations were tedious to do by hand—but that changed in the twentieth century with the invention of the computer. Bayes's theorem was adopted by insurance companies, the military, and the technology industry. It is no exaggeration to say that the Reverend Bayes's long-forgotten rule is behind much of Silicon Valley's wealth.

"If you're not paying for it, you're the product being sold." This is a maxim of our digital economy. Google, Facebook, Instagram, Twitter, YouTube—all our entrancing and addictive apps—are free products that come with a Faustian bargain. To use these services we allow their providers to collect so-called personal information—information that is valuable because of Bayes's theorem. In the aggregate, as "big data," personal information allows marketers to predict what you will buy, how much you will pay, and whom you will vote for. These Bayesian predictions, updated with every click, swipe, post, or GPS coordinate, are the secret sauce of many a tech company.

This success story is, however, only the prologue to the stranger one that concerns us. In recent years it has been recognized that Bayesian methods can shed light on deep mysteries of existence, including the future of the human race itself.

Ozymandias

I met a traveller from an antique land
Who said— "Two vast and trunkless legs of stone
Stand in the desert.... Near them, on the sand,
Half sunk a shattered visage lies, whose frown,
And wrinkled lip, and sneer of cold command,
Tell that its sculptor well those passions read
Which yet survive, stamped on these lifeless things,
The hand that mocked them and the heart that fed;
And on the pedestal these words appear:
My name is Ozymandias, King of Kings;
Look on my Works, ye Mighty, and despair!
Nothing beside remains. Round the decay
Of that colossal Wreck, boundless and bare
The lone and level sands stretch far away."

This is the sonnet "Ozymandias" (1818) by Romantic poet Percy Bysshe Shelley, husband of *Frankenstein* author Mary Shelley, daughter of feminist Mary Wollstonecraft, friend of minister Richard Price, promoter of the intellectual property of Thomas Bayes. The theme of "Ozymandias" is that glory is fleeting. Nothing lasts.

In the summer of 1969, J. Richard Gott III celebrated his Harvard graduation with a tour of Europe. He visited the supreme monument of Cold War anxiety, the Berlin Wall. Standing in the shadow of the landmark, he contemplated its history and future. Would this symbol of totalitarian power one day lie in ruins?

This was a matter discussed by diplomats, historians, op-ed writers, TV pundits, and spy novelists. Opinions varied. Gott, who was planning postgraduate work in astrophysics, brought a different perspective. He devised a simple trick for estimating how long the Berlin Wall would stand. He did the math in his head and announced his prediction to a friend, Chuck Allen. The wall would stand at least two and two-thirds more years but no more than twenty-four more years, he said.

Gott went back to America. In 1987 President Ronald Reagan demanded, "Mr. Gorbachev, tear down this wall!" From 1990 to 1992 the wall was demolished. That was twenty-one to twenty-three years after Gott's prediction and within the range he announced.

Gott called his secret the "delta *t* argument." "Delta *t*" means change in time. It's also known as the *Copernican method,* after Nicolaus Copernicus, the great Polish astronomer of the Renaissance. Copernicus's leap of imagination was that the Earth is not the center of the universe. It is only one of a number of planets circling the sun. This thinking led to a simpler model of the solar system, one that agreed better with observation.

To astronomers, Copernicus's insight has been a gift that keeps on giving. Over the past five centuries it has been established again and again that humanity does not occupy a central or special place in the scheme of things. Our sun is an ordinary star in an ordinary galaxy. It is not at the center of the galaxy but well off to the margins. Our galaxy does not occupy a special place in the cluster of galaxies to which it belongs, and this cluster has no special place in the universe as we know it. Even the whole of the observable universe is now widely believed to be an insignificant speck in a yet-greater multiverse. The cosmic "you are here" dot says we're smack in the middle of nowhere.

The Copernican principle is generally applied to an observer's location in space, but the delta *t* argument applies it to an observer's location in time. Gott began with the assumption that his visit to the Berlin Wall had not taken place at any special moment in the wall's history. That premise allowed Gott to predict the wall's future without any expertise on Cold War geopolitics. His 1969 prediction was that there was a 50 percent chance that the wall would stand at least another 2.67 years after his visit but no more than 24 years.

Gott published his method in the prestigious journal *Nature* in 1993, and it ignited a controversy that still burns white hot. Many insisted that Gott's method could not possibly be valid. They cited erudite (and remarkably different) reasons. Some discerned in Gott's article a symptom of a jaded intellectual culture. "In the age of Quantum Mechanics, we often embrace a fantastic conclusion simply because it is fantastic and shocking," complained George F. Sowers Jr. "Our sensibilities have been numbed. But the world is not so topsy-turvy that we can reason *à la* doomsday."

Still others reported that they had tried Gott's method, and it worked. A group of British mathematicians used Gott's idea to compute how much longer the Conservative Party would remain in power. In line with their prediction, the party was ousted three and a half years later.

How Long Will Love Last?

Gott is a colorful character, literally. When I met him he was wearing a turquoise jacket of almost fluorescent hue and a tan fedora. He is a natural storyteller, with a Kentucky twang that has survived decades in the Ivy League, and a droll sense of humor. In the years after the appearance of his *Nature* article he became a minor celebrity as a sort of scientific soothsayer. In 1997 Gott invited readers of *New Scientist* to use the arrival time of the magazine to estimate how long they would be with their present boyfriend, girlfriend, or spouse. The principle can apply just as well to readers of this book.

You are now reading these words at a random moment in the course of your romantic relationship. It can hardly be otherwise. This isn't a book about how to tell if he or she is really into you. It's not a book about how to find a good divorce attorney. This book might have come into your life at almost any time. That's the unromantic Copernican assumption. *There is nothing at all special about this moment.*

Chances are, then, that you are not at the very beginning of the relationship, nor at the very end. You're somewhere in the middle. If you accept this premise, the past duration of your relationship gives a very, very rough idea of its future duration.

You may recognize this as common sense. If you met someone five days ago, it wouldn't be surprising for the affair to be over five days from now. It's too early for a tattoo or a deposit on a beach house for next summer. You may find this kind of estimation amusing or depressing or both. But the real question is, how accurate should we expect such estimates to be?

Gott realized that you don't need fancy math to calculate that. All it takes is a diagram you can sketch on a napkin.

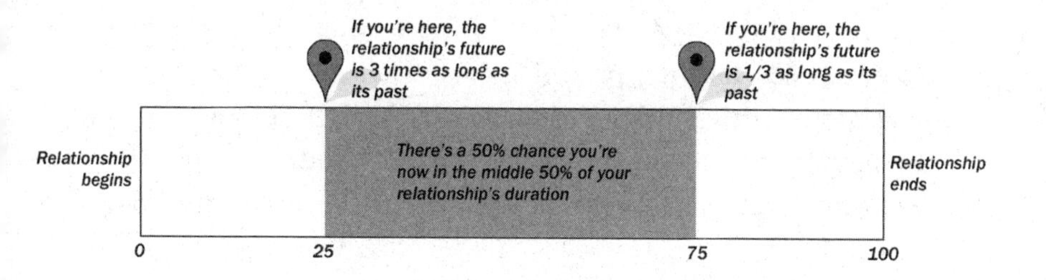

Draw a horizontal bar representing your love affair's duration in time. Think of it as the scroll bar of a movie. The relationship's beginning is at the left, and its end is at the right. Since no one knows how long love will last, we can't mark the bar in hours, days, or years. Instead we'll mark it in percentage points. The relationship's beginning is at 0 percent, and its end is at 100 percent (however long that is in real time). The present moment must fall somewhere between 0 and 100 percent, but we don't know where.

Still with me?

I have shaded half the bar. It's the middle half, running from 25 to 75 percent. The present moment can be represented by a map pin ("You are here"). We'll assume it's equally likely to fall anywhere along the bar's length. That could be in the shaded part or the unshaded part. But because the shaded region is exactly 50 percent of the bar, we can say that the odds are 50:50 that the current moment falls within the shaded part.

I've put two sample pins on the diagram. They mark the ends of the shaded region. The left pin is at 25 percent. There is no reason to believe that this pin corresponds to where you are in your relationship's timeline. But suppose for the sake of argument that it does. Then your love has lasted 25 percent of its total duration, and it still has another 75 percent to go. The future is three times longer than the past.

The pin on the right is at 75 percent. Should that be the correct position, the future (the 25 percent remaining) is only one-third as long as the past (75 percent).

Because these two pins bound the middle half of the bar, it's even odds that the present moment falls inside this range. That means there's a 50 percent chance that your relationship's future will be somewhere between one-third and three times as long as its past. Gott used this calculation with his Berlin Wall prediction.

This prediction is one of many similar ones you might make. In his *Nature* article, Gott adopted the 95 percent confidence level that is widely used in science and statistics. To publish a result in a scientific journal, it is generally necessary to show a 95 percent or greater probability that the result is not due to sampling error. You don't have to be a scientist to appreciate that 95 percent is pretty confident. Is this Mr. Right or is it just Mr. Right Now? You're never 95 percent sure of that. Nor are you often 95 percent confident of tomorrow's weather or the winner of the next election.

I've made another diagram with the middle 95 percent of the bar shaded. This time the shaded region runs from 2.5 to 97.5 percent. Should you find yourself at the left pin, you have 2.5 percent of the duration behind you and 97.5 percent ahead. The future is 97.5/2.5 or 39 times as long as the past.

At the right pin, the future is only 1/39 as long as the past. Thus the range for 95 percent confidence, in this or any other Copernican estimate of future duration, is 1/39 to 39 times the past duration.

$$\text{past time}/39 < \text{future time} < \text{past time}*39$$

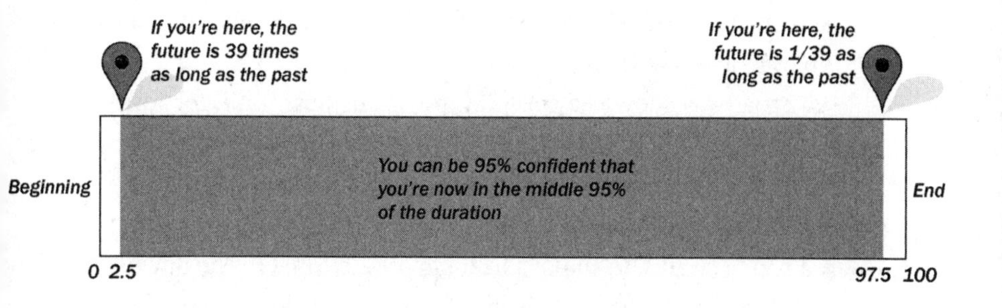

For example, let's say you met someone a month ago. You can be 95 percent confident that this relationship will end in no less than 1/39 month and no more than 39 months. That spans about eighteen hours to a little over three years. You can be reasonably sure you won't miss a break-up text when you switch off your ringer for a movie. You should also expect that you won't be involved with this person five years from now—so say Gott's statistics of love.

Lindy's Law

Over the years, Gott and others have claimed diverse applications of the Copernican method. Take Wall Street's famous weasel words: past performance is no guarantee of future results. Nonetheless an incredible amount of effort goes into divining future stock performance from (what else?) the past.

Statistics on corporate survival—and on tenure on ranked lists or indexes like the Fortune 500 or the S&P 500—show a Copernican effect. How long a company has existed (or been on the ranked list) is a rough predictor of how long it will survive (remain on the list).

The Copernican principle has some relation to the survivor bias that plagues stock investors. At any given time, an index fund or portfolio tends to be weighted with stocks that have done well in the immediate past, but that are unlikely to perform comparably well in the long run. Investors are always grabbing gold that crumbles to ashes in their hands.

A Broadway show is a special type of business. Like corporations, plays run for as long as their investors can hope to make a profit. But compared to corporations, Broadway shows are mayflies, with lifespans measured in weeks. Gott realized that that offered him a chance to make a testable prediction. On the day his 1993 *Nature* article was published, he identified forty-four plays and musicals that were then running in New York, including hits like *Cats* as well as productions that were quickly forgotten. Four years later thirty-six of the forty-four plays had closed, all within Gott's prescribed 95 percent confidence intervals.

It was recently reported that 79 percent of Broadway musicals are flops, closing before they recoup their costs. Tax write-offs notwithstanding, it appears that many backers of plays overestimate runs. Gott's prediction method does not factor in playwright, stars, casts, or reviews; nor does it consider advance ticket sales, celebrity buzz, advertising campaigns, or what people are willing to pay or do to score a ticket. He nonetheless found that how long a show had already run was a better predictor of its future run than much informed opinion is. *The New Yorker*'s editors were impressed enough with Gott and his methods that they commissioned Timothy Ferris to write a profile of Gott. The 1999 article ran with the title "How to Predict Everything."

A grad student gazes, Zen-like, at a wall and gains enlightenment. Can it really be that easy to predict "everything"?

Self-Locating Information

It certainly seems that Gott's method pulls a big, dramatic prediction out of an empty hat. But you can't conjure a prediction out of nothing. In fact, the Copernican method uses a special kind of information.

An example is that embodied in the Google map pin. Designed by Jens Eilstrup Rasmussen in 2005, the upside-down teardrop quickly became a global shorthand, earning a place in the Museum of Modern Art's design collection. Rasmussen's icon epitomizes the power of digital media over print. Search all the world's printed road maps and atlases. Never will you find the most important information you can get from a map—where you are and where you're going.

The digital map user is never lost. That's because a GPS-enabled map has something extra. It knows where the user is. This is *self-locating* or *indexical* information. Those are fancy terms for something we largely take for granted. "Indexical" refers to the index finger, pointing to someone or something. "You are here."

Self-locating information need not pertain to a position in space. It can also describe a location in time. This too can be useful. (Otherwise why would we have clocks?) Gott's Copernican method uses one's position in time to make its predictions.

Forecasts from self-locating information are nothing new. In 1964 biographer and critic Albert Goldman formulated "Lindy's law." "According to a law established and promulgated by bald-headed, cigar-chomping know-it-alls who foregather every night at [New York deli] Lindy's...the life expectancy of a television comedian is proportional to the total amount of his exposure on the medium." Many comics who score a *Tonight Show* shot are soon forgotten, but it's safe to assume that Jerry Seinfeld will be around awhile.

Mathematician Benoit Mandelbrot came across Lindy's law and wrote about it, saying that it applies to many things other than show business. That was Gott's point.

Before I heard of the Copernican method, I formulated a semiserious law for waiting on hold to speak to a customer support agent. Your future wait to speak to a live human is approximately equal to however long you've already waited. It is only in the first few seconds of being on hold that you may cherish the prospect of speaking to an agent right away. As the seconds turn to minutes, so does your expected wait time.

Copernican estimates need not strictly involve a duration. Say you're flipping channels and come across a movie titled *Rocky IV*. How many Rocky movies did they make, anyway? If this random one is number four, "about eight" is a decent guess.

There is some fine print to this form of divination. Gott put it this way: you can't entertain wedding guests with uncanny forecasts of the newlyweds' future breakup. Not only would that be gauche; it wouldn't work.

The method is grounded in the premise that you find yourself at a random point in the duration of something. We can't hop in a time machine and set it for "random." In practice this means that you have to be in a situation where you have no way of knowing where the present moment falls within the duration of the phenomenon of concern, and no reason to believe the present moment is early, in the middle, or late.

A wedding is a celebration of the beginning of a shared life. All hope it's an early moment in the relationship, not a random one.

Then there are longevity effects. At Frank and Fran's fiftieth anniversary, someone wishes them another fifty years of wedded bliss. That's a joke, not a prediction. We can infer that Frank and Fran are in it for the long haul, but that doesn't override what we know about human lifespans.

Riddle of the Sphinx

I must now tell how Gott made his doomsday calculation, concluding that the final curtain of that show in which we are all actors may fall sooner than we think.

Cynics may ask, *What else is new?* It's not difficult to find cause for pessimism in the day's news. But Gott came to this determination from a different direction. It was all math, taking no account of whatever common knowledge we have about war, terrorism, environmental disaster, out-of-control technology, and other specific threats to human life.

In his 1993 *Nature* article, Gott laid out what is now called the doomsday argument. He described two versions of it, one using our point in time, and another, developed by astrophysicist Brandon Carter and philosopher John Leslie, using our position in a chronological list of human beings. Either way, the doomsday argument predicts a date for the extinction of the human race.

Archaeologists say the first anatomically modern human remains date from about 200,000 years ago. Skulls of that time enclosed brains about the size and shape of ours. Suppose then that we're at a random point in the timeline of human existence. There are 200,000 years behind us, and we can expect something like 200,000 more years ahead of us—very, very roughly. Using 95 percent confidence levels, Gott estimated that the human race would survive at least 5,100 more years but no more than 7.8 million years.

Biologists have put the average lifespan of mammalian species at 1 to 2 million years. Gott's range is consistent with that and shouldn't be regarded as gloomy. Note that this prediction says there is only a 2.5 percent chance of human extinction in the next 5,100 years. It's easy to feel that this is too optimistic.

But there are other, finer-grained ways of looking at it. Right now is *not* such a random point in the existence of the human race. The best way to demonstrate that is with a chart of world population over time.

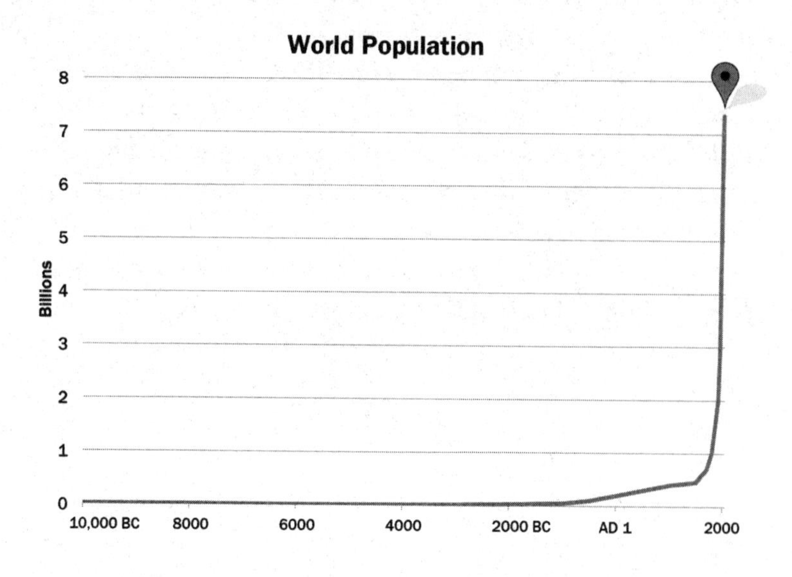

The chart is a hockey-stick curve. World population burgeoned with the adoption of agriculture, metalworking, industry, and digital technology. Everyone whose name we know, from Homer to Taylor Swift, is crowded into the most recent 1.5 percent of the timeline of our species. The map pin shows the present moment, and it's clearly not typical.

"Or consider lemmings," said John Leslie. "Where does a typical lemming find itself? At a time when there are hardly any other lemmings, or after a lemming population explosion?" (Lemmings are arctic rodents whose population varies widely in multiyear cycles. A myth holds that they commit mass suicide by leaping off cliffs into the ocean.)

So we live during a population spike. Carter and Leslie's approach has a way to accommodate that. It uses a technique called *self-sampling*. Let's say I buy a ticket to a raffle, and it's number 64. That gives me some conception of how many tickets were distributed. Assuming consecutive numbering, there have to be at least 64 tickets. But there probably aren't millions, as I would have been unlikely to get such a low number as 64.

In self-sampling you regard yourself as a random sample of a group. You then use knowledge about yourself (like your ticket number) to draw conclusions about the group (the total number of ticket holders). In their version of the doomsday argument, Carter and Leslie use birth order.

Imagine a list of every past, present, and future person, sorted by year and time of birth. Conceptually the list would be like this:

1. Adam
2. Eve

...

X. Me

...

Z. The last human being (from the future)

My birth-order serial number is *X*. Am I near the top of the list, toward the middle, or near the bottom? How long is the list, anyway?

I don't know that. I can say only that there is no reason to believe my position in line is too atypical. This is again the Copernican assumption, only now with birth ranks rather than years. We're using the ticks of a "birth clock" rather than those of a regular clock.

Gott cited estimates putting the cumulative human population at about 70 billion. This is everyone who ever lived, up to the present. The number is better defined than you might think. Prehistoric populations were tiny by today's standards. It therefore doesn't matter too much where you draw the arbitrary line between *Homo heidelbergensis* and early *Homo sapiens,* or whether you count human-Neanderthal hybrids. These early peoples wouldn't contribute much to the total head count.

Because nearly all lives cluster to the rightmost part of the population chart, we have documents and archaeology to help estimate the populations of recent millennia.

My birth serial number would be somewhere around 70 billion. That's *X.* But what about *Z,* the birth rank of the last human?

With 95 percent confidence it would be between 1/39 and 39 times my number. Gott estimated the number of people yet to be born at 1.8 billion to 2.7 trillion.

Now we need a way to convert future births into years. How many years will it take for the Omega Person to be born?

That depends on the birth rate. At the time of Gott's article, there were about 150 million births a year. If that rate were to continue it would take only 12 years to add another 1.8 billion people. Attaining the higher limit of 2.7 trillion future births would take 18,000 years.

By this calculation we would expect doomsday to fall somewhere between 12 years and 18,000 years from now. This is an alarming projection—especially that lower limit.

The world's birth rate has declined a bit since 1993 (to about 130 million a year). Meanwhile estimates of the cumulative population now tend to be larger than the figure Gott used (somewhere around 100 billion). These revised numbers bump the range up to 20 years to 30,000 years from now. We still can't confidently exclude doomsday happening within the natural lifespan of people living today.

We really ought to use the *future* birth rate. That is an unknown. One possibility is that the human race continues to grow more or less exponentially. This is conceivable in a future in which humans occupy other planets, or one in which technology supports population densities presently inconceivable on Earth. With this assumption we would reach the birth-order milestone Z all the sooner. It would accelerate doomsday, not put it off.

A seemingly more benign assumption is that the birth rate continues to decrease. But to defer doomsday significantly, we'd need drastically fewer births. It's hard to put a positive spin on that. It might entail a global catastrophe leaving a few postapocalyptic survivors. Were the number of births per year to drop by a factor of one hundred, this could put off human extinction by a factor of one hundred. It takes near-doomsday to put off total doomsday. That's not much of a victory.

Magic 8 Ball

It is the early hours of August 7, 2015. You're sitting in Lindy's reading the first reviews of a musical that opened the previous night.

> I am loath to tell people to mortgage their houses and lease their children to acquire tickets to a hit Broadway show. But "Hamilton," directed by Thomas Kail and starring Mr. [Lin-Manuel] Miranda, might just about be worth it—at least to

> anyone who wants proof that the American musical is not only surviving but also evolving in ways that should allow it to thrive and transmogrify in years to come.
>
> —*Ben Brantley,* New York Times

There are limits to Gott's brand of prognostication. A rave review by an influential critic can be a legitimate reason for believing a show will run for a long time—however brief its run has been so far. It's possible to feel the same applies to us. *Homo sapiens* have survived mammoths and malaria and atom bombs. Nothing's killed us yet. We are no typical species, and this is no typical moment—so get over it.

In other words, we might have a strong conviction that the human species will survive a long time. This belief is a prior probability. Certainly, the end of the world is a matter on which opinions vary. Some cultists and pessimists are sure the end is near. Some optimists are convinced humans will survive for billions of years. There are Nostradamus-spouting psychics who claim to know the exact day, hour, and minute of doomsday (and can forecast your love life more precisely than any math).

Gott did not mention prior probabilities in his 1993 paper. But they are central to still another version of the doomsday argument, developed by Carter and Leslie. It uses Bayes's theorem to adjust prior probabilities for the new evidence supplied by birth rank.

Unfortunately this third doomsday prediction does not supply an easy reprieve. Given almost any reasonable optimism about the future, the Bayesian doomsday argument shifts the odds to end up with a high probability of impending catastrophe. Carter has described his version of the doomsday argument as a magnifying glass. It says that the probability of doomsday is bigger than you thought it was. Apocalypse is closer than it appears in the mirror.

I'll give a simplified model in which there are only two possible scenarios: "doom soon" and "doom later." Doom soon means that humans will become extinct within five hundred years. Doom later means we will survive beyond that, achieving a cumulative population a thousand times greater than it would have been with doom soon.

Let's say I begin with the belief that the chance of doom soon is 10 percent. In this toy example Bayes's rule shifts that chance upward to about *99 percent.* (For those interested in the math, see end note on page 266.)

Maybe I'm more optimistic and believe the chance of doom soon is only 1 percent. Bayes pushes that upward to 91 percent.

A super-optimist might think the chance of doom soon is only 0.1 percent. Bayes raises that to 50 percent.

Over virtually the full spectrum of rational beliefs about the future, early human extinction ends up being more likely than not—*if* the Bayesian doomsday argument is valid. Not since Malthus has a demographic forecast inspired such intense controversy. Will we resolve our differences, banish war and terrorism, save the environment, and go on to explore the galaxy? Bayes's Magic 8 Ball says, *VERY DOUBTFUL.*

Riddle of the Sphinx

The entertainment industry mints franchises out of doomsday (or the threat thereof, evaded in the nick of time). There are cinematic genres built around nuclear war, asteroids headed for Earth, villains bent on global destruction, zombie marauders, robot usurpers, and extraterrestrial invaders. It is impossible to live in our culture without being exposed to the idea that species, no less than individuals, can be mortal. Memento mori. *Hasta la vista*, baby.

The doomsday argument is a different kind of premonition.

Its oracle is maddeningly silent on what will extinguish human life.

In the not-too-distant past, any who felt the end was near would likely have assumed nuclear war as the cause. Today the list of existential threats is longer, and artificial intelligence (AI) rivals the bomb as a disturber of sleep.

It is ironic that an undercurrent of pessimism pervades Silicon Valley, those few golden square miles that, more than any other part of the globe, have been enriched by Bayes's theorem. Much of the ambivalence about AI has its roots in the work of Norwegian-born philosopher Nick Bostrom, now of Oxford. Bostrom did his doctoral thesis on the doomsday argument and the puzzles of self-sampling. He has been influential in proposing that self-sampling can be applied to diverse scientific questions. Today Bostrom is largely concerned with the risks that may be posed by AI. He believes that the challenge of coding human values into machines is more formidable than is generally appreciated. AI could one day be all-powerful. Getting it wrong would be catastrophic.

This book will trace the remarkable though little-heralded intellectual adventure that began with doomsday. By applying Thomas Bayes's rule to the technique of self-sampling, we can address cosmic mysteries. Was life on Earth probable or a rare accident? Why don't we see any evidence of extraterrestrials? Is the world we see real or a simulation? Is the universe we observe all there is?

It's little wonder that, in just a few years, the doomsday argument has become a pivot of contemporary thought. It is that rare philosophical dispute that offers an accessible gut punch of a premise. It not only links to trending topics in science, technology, and culture; it also has the potential to help answer big questions of life, mind, and the universe. The doomsday argument is the sphinx's riddle of our age, and we're playing for life and death.

The Minister of Tunbridge Wells

I never saw a worse collection of human creatures in all my life." That was Elizabeth Montagu's 1745 assessment of the Kentish spa town of Tunbridge Wells, a resort then drawing polyglot aristocrats and social climbers from throughout Europe. Montagu, the London hostess and bluestocking, later tempered her opinion, allowing that "the variety of persons and characters make Tunbridge an epitome of the world."

Today Tunbridge Wells is known as Jane Austen country. Jane's father, the Reverend George Austen, spent his boyhood in the area. The town figured in the imagination of the Austen family as its fortunes dwindled. Lately Tunbridge Wells's own fortunes have improved because of its association with the Austens. Mentioned in several of Jane's novels, the town has become a pilgrimage site for Austen fans and a location for film adaptations.

There is also an E. M. Forster connection. "I am used to Tunbridge Wells, where we are all hopelessly behind the times," sighs Lucy Bartlett in *A Room with a View* (1908). By Forster's time, the faded resort was being pegged as an emblem of ossified British conservatism. Since the 1940s, "Disgusted of Tunbridge Wells" has been a facetious pseudonym for letters to the editor expressing stodgy views.

Tunbridge Wells was nonetheless the birthplace of one of the contemporary world's most disruptive ideas. Not many traces remain of the town's onetime minister Thomas Bayes (1701–1761). The Bayes family had made its fortune several generations earlier, in the cutlery business of Sheffield. Thomas Bayes studied theology and logic at the University of Edinburgh. After several years in London, he moved to Tunbridge Wells in 1733 or 1734 and became minister of Mount Sion Chapel. Bayes was a Presbyterian nonconformist, opposing the Church of England and the *Book of Common Prayer* on grounds vague to nearly all of today's Presbyterians.

Bayes did not achieve renown for his sermons, and he was so obscure that there is no known portrait of him. Yet he secured a connection to London's scientific circles. The second Earl of Stanhope, a dilettante mathematician with a country seat near Tunbridge Wells, had Bayes inducted into the Royal Society. Stanhope was impressed by an article Bayes had written defending Newton's calculus against the criticisms of Bishop Berkeley. It was one of the two articles Bayes published in his lifetime. The other was a work of theology titled "Divine Benevolence, or an Attempt to Prove That the Principal End of the Divine Providence and Government Is the Happiness of His Creatures."

The great minds of the Enlightenment were discarding Church teachings right and left. Scottish philosopher David Hume's *An Enquiry Concerning Human Understanding* (1748) ignited an eighteenth-century culture war by questioning the reality of Christian miracles. The Bible says that Jesus walked on water, turned water into wine, multiplied loaves and fishes, raised Lazarus from the dead, and returned from the dead himself. Hume boldly proposed that the standards of evidence applying in a court of law ought to apply to miracles. Hume favored the Scottish verdict of "not proven."

The thing about miracles is that they happen once and can't be repeated for Doubting Thomases. You had to be there, and probably you weren't. Hume argued that it is fitting to consider both the intrinsic probability of an event and the credibility of the testimony establishing it: "No testimony is sufficient to establish a miracle, unless the testimony be of such a kind, that its falsehood would be more miraculous than the fact which it endeavors to establish."

As both a mathematician and a clergyman, Bayes must have felt himself in the line of fire. He would have had reason to ponder how, and whether, belief in miracles could be reconciled with the Enlightenment. It is conjectured that Bayes's work on probability was motivated by Hume's debunking of miracles. But there is no mention of Hume or miracles in Bayes's one influential work, the one describing his theorem: *An Essay Towards Solving a Problem in the Doctrine of Chances*. Nor do we know for certain when the *Essay* was written. Richard Price discovered it after Bayes's death, filed among papers of the late 1740s.

Bayes's Theorem

The theory of probability began at the gambling table. Gerolamo Cardano was the ultimate Renaissance man—a philosopher, mathematician, physicist, astronomer, astrologer, inventor, chemist, biologist, and fashionable physician. He was also a compulsive gambler who by his admission bet daily for twenty-five years. Cardano's short treatise on probability was an attempt to understand how so much money had slipped through his fingers. Gamblers already knew how cards, dice, and roulette wheels worked. They needed to know the odds: how to calculate the chance of drawing two aces, rolling a 7, or winning repeated bets on red. Cardano and his French successors Pierre de Fermat and Blaise Pascal supplied that long before Bayes's time.

Bayes took up the opposite issue: *inverse probability* or the *probability of causes.* Suppose we already know the outcomes (the hands we've been dealt). What can we conclude about the causes (whether the dealer is honest or a cheat)? This too is a pressing question for any serious gambler.

Should the dealer be using sleight of hand to avoid giving out aces, that would affect the hands I receive. Bayes's rule provides a mathematical framework for reasoning about such matters. It starts with a prior probability, such as "The chance of drawing an ace from a fair deck is 1/13." Each card dealt allows me to adjust this probability up or down, to reflect the changing composition of the deck and my ongoing experience with the dealer. The adjustment produces a *posterior probability,* updating the prior for the new evidence.

Should I find that I'm consistently drawing less than my fair share of aces, I can infer a cause—a cheating dealer or a deck that's missing an ace. This inference is never 100 percent certain. It remains conceivable that I'm having a terrible run of bad luck. But the probability of cheating increases as the "bad luck" continues. We live in a world where nothing is certain. The reasonable gambler must walk away from a game that is probably rigged.

The *Essay* is distinguished as one of the worst mathematical papers describing a great concept. Bayes's exposition is now judged to be flawed, confusing, and unresolved—and littered with analogies that are harder to understand than the points they attempt to clarify. Price's introduction adds a spin that Bayes himself did not supply. Price frames the *Essay* as a dog whistle for believers in the ultimate cause, the Christian God: "The purpose I mean is, to shew what reason we have for believing that... the world must be the effect of the wisdom and power of an intelligent cause; and thus to confirm the argument taken from final causes for the existence of the Deity."

Price's sentiment is what we today call an argument from design. The universe is a beautifully constructed watch, from which we can infer a divine watchmaker.

Bayes's *Essay,* however, is strictly a work of math. Its thesis is in many ways commonsensical. Let's begin by walking through some elevator pitches for the Bayesian philosophy.

1. "Extraordinary claims require extraordinary proof." This five-word adage of contemporary skeptics is not a bad introduction to Bayesian thinking. To use Hume's example, the Bible says that Jesus was the son of a carpenter, that at an early age he impressed elders with his wisdom, that he gave a sermon on a mount, and that he had a supper with followers before he was crucified on the order of Pontius Pilate. The New Testament is the only source for these assertions. They are almost universally accepted as true. It is rather the New Testament's miracles that are denied by non-Christians. Why? It's one thing to say that the four evangelists might be unreliable narrators. If so, shouldn't all biblical events be equally suspect?

Not necessarily. Miracles are extraordinary claims demanding a higher bar of proof. One-shot miracles have a low prior probability, based on everything else known about how the world works. The scriptural evidence (of being asserted in an ancient text that appears to combine biography, legend, and allegory) is insufficient to raise that probability very much. But incidental details such as being the son of a carpenter start with a much higher probability of being true. There the biblical account is sufficient to boost that chance to likelihood, even for a nonbeliever.

Bayes and Price were men of faith. Price's writings suggest he saw Bayes's theorem as a holy loophole, allowing Enlightenment Christians to preserve their belief in miracles. If enough witnesses attested to a miracle, each observation could incrementally elevate its probability to near-certainty.

This demonstrates one common complaint about Bayes's theorem, that it leaves much to the judgment of the user. It does, and the same may be said of all rules, laws, and credos ever applied by fallible mortals.

2. Absence of evidence can be informative. In Arthur Conan Doyle's "The Adventure of Silver Blaze" (1892), Sherlock Holmes is investigating the murder of a horse trainer. The detective notices that none of the witnesses mentioned hearing the stable's watchdog barking. The dog would have barked had the villain been a stranger. Holmes deduces the murderer was someone known to the victim and the dog.

Doyle joins Bayes in making the subversive point that the lack of evidence (a dog *not* barking) can be as revealing as affirmative evidence is. Bayes's rule says to look at the ratio of probabilities. The dog not barking is probable with a familiar visitor but improbable with a stranger. That is reason to favor the first possibility.

3. "When you hear hoofbeats look for horses, not zebras." All else being equal, the more common explanation is to be preferred.

Here's another example: In the third grade I won a trophy for kickball. Which was more likely?

- I won the trophy because I was the best at kickball out of all the kids in the third grade.
- I won because it was a participation trophy (handed out to every kid to boost self-esteem).

The second makes my win a given rather than a dramatic victory against the odds. That's reason to think the second hypothesis more likely. "Don't assume an observation is extraordinary when it could easily be regarded as ordinary," as John Leslie put it. We should not be too ready to attribute our reality to flukes, long shots, and weird coincidences.

Homer Simpson and the Urn

The Springfield County Fair is offering a game of skill and chance. It involves two large, identical urns. One contains ten balls and the other one thousand, but neither urn is labeled. The balls in each urn are numbered consecutively—number 1 to number 10 in one and number 1 to number 1,000 in the other. The customer picks an urn, and the operator draws a single ball from it, revealing its number. The customer must then guess how many balls are in the chosen urn to win a Kewpie doll.

Homer Simpson puts down a dollar to play. He picks the left urn.

The operator draws a random ball from the left urn. It's number 7. "All right, pal. How many balls in this urn?"

"A thousand!" guesses Homer.

Poor Homer has failed to apply Bayes's theorem. Before the ball's drawing there was no reason to think that either urn was more likely to be the one with a thousand balls. The likelihood was 50:50 for each. The random sampling reveals new information that Homer ought to use. Drawing the low number 7 from an urn greatly increases the chance that it contains *only* low numbers.

If the left urn has just ten balls, the chance of drawing number 7 from it would be 1 in 10. If it has one thousand balls, the chance would be 1 in 1,000. Drawing number 7 is not a likely outcome in either case. But we already know that number 7 has, in fact, been drawn. Common sense tells us the sampled urn is almost certain to hold ten balls. It further hints that the odds ought to be 1,000:10, or 100:1, in favor of the sampled urn having just ten balls. These are exactly the odds you get by applying Bayes's theorem.

I'll now give a simple statement of Bayes's theorem. You've

probably heard of false positives and false negatives. Medical tests can give an accurate result, indicating that a patient has a certain condition when he does (a true positive), or a misleading result, saying he has the condition when he doesn't (a false positive). This terminology provides a concise way of expressing Bayes's theorem. The chance of something, after a "test," is the chance of a *true* positive result divided by the chance of *all* positive results (true and false).

Should you prefer an equation, here goes:

$$P(H|E) = P(H\&E)/P(E)$$

$P(H|E)$ is the probability we want to calculate. It is the chance of a hypothesis (such as "This urn has ten balls") being true, *given* that I have obtained some evidence bearing on it (such as drawing a low-numbered ball). Bayes's rule says that this probability is equal to $P(H\&E)$, the chance that the hypothesis is true and the evidence supports it (a true positive), divided by $P(E)$, the total chance of obtaining that evidence (either as a true positive or a false positive).

Apply that to the Springfield urns. We want to test whether an urn has just ten balls. Drawing a low number, 1 through 10, is a positive test result implying that it is likely that the urn has ten balls. A true positive is drawing a low-numbered ball from the urn that actually does have ten balls. The chance of this occurring is 50 percent.

That's because there's a 50 percent chance of picking the ten-ball urn for the drawing. Should you do that, you're guaranteed to draw a ball numbered no higher than 10, and this result must be a true positive. (Should you pick the thousand-ball urn, you can't have a true positive for the ten-ball test regardless of what number you draw.)

The chance of *all* positives is the chance of a true positive

(50 percent, we just decided) plus the chance of a false positive. To get a false positive, you'd have to pick the wrong, thousand-ball urn (50 percent chance) and also happen to misleadingly draw a low-numbered ball from that urn. The chance of that is only 10 out of 1,000, or 1 percent. That means the chance of a false positive is 50 percent times 1 percent, or 0.5 percent.

To sum up, Bayes's theorem says that the chance an urn has ten balls, given that you drew a low-numbered ball from it, is 50 percent/(50 percent + 0.5 percent). This comes to 100/101 (100:1 odds) or just over 99 percent. Homer should be very confident the left urn has just ten balls.

Nothing here is mathematically deep or clever. It's just commonsense accounting. Homer's problem may be that he figures the drawing of a number 7 doesn't tell him anything. Had the chosen ball been number 11 or higher, then he could have deduced, with ironclad certainty, that the draw was from the urn with a thousand balls. But there's a ball number 7 in both urns. The evidence supplied by drawing a number 7 ball is circumstantial but not to be neglected by any fully reasonable party.

"Rational belief is constrained," Nick Bostrom wrote, "not only by chains of deduction but also by the rubber bands of probabilistic inference."

Doubting Thomas

Bayes's *Essay* found a most influential reader. He was Pierre-Simon Laplace (1749–1827), a French marquis, mathematician, physicist, astronomer, and atheist. Laplace slapped Bayes's wreck of a paper into rigorous math. There are those who judge Laplace to be the true originator of Bayesian probability, leaving Bayes as simply a brand.

Everyone read Laplace. But Laplace's enthusiasm for the probability of causes couldn't change reality. In the simplest instances

Bayes's result was obvious even without the math. In others the subjective nature of prior probabilities made it hard to decide who or what was right. In still other cases it was difficult to do extended calculations by pen and ink. Anyone attempting to update probabilities repeatedly risks losing patience before gaining much insight.

Over the following centuries, the disciplines of probability theory and statistics took a different path. Most scientific observations are not one-off miracles. They can be repeated at will, and a good scientist is expected to play Doubting Thomas. Perform the same experiment the same way, and the same result should obtain in London, Lucknow, or Lima. When a result can't be repeated, that raises a red flag.

We shouldn't pay too much attention to anecdotal evidence. Everyone's got a neighbor, coworker, or friend-of-a-friend who's been helped by some novel supplement or regimen. The way to find out whether a treatment works is to do a randomized, double-blind trial. If the treatment really works, the people getting it should do better than those who get a placebo, and the difference must exceed expected bounds of statistical error.

Repeatability and random trials rank with the greatest monuments of modern thought. Much of modern statistics has explored how best to design experiments or sample populations, and how to interpret the data. This focus marginalized Bayesian probability, at least until the twentieth century and its calculating machines.

No one really knows what Thomas Bayes thought his theorem was good for. He couldn't have anticipated the diverse uses to which it has been put. Bayes's rule has fought Nazis and internet spammers.

When Allied forces were planning the D-Day invasion, they needed to estimate how many Panzer V tanks the Germans were producing. The Allies had captured a number of German tanks. They knew that German manufacturers were meticulous about serial numbering. There were serial numbers on tank gearboxes, engines, and chassis. Since the captured tanks could be regarded as random drawings from the whole set of existing tanks, this allowed military statisticians to estimate the German tank production. They put it at 270 tanks a month, a fraction of what spy data had claimed. Records uncovered after the war showed the Germans had been producing 276 tanks a month. The statisticians' estimate was almost on the nose.

Today, so-called Bayesian spam filters use a continuously updated list of words and phrases that appear in unwanted email messages. A typical list might include *FREE, Earn $, cure baldness, Viagra, dig up dirt on friends, work from home, score with babes, score with dudes, you are a winner!* None of these words or phrases can prove a message is spam. This paragraph contains all of them, and it isn't spam. But a message having one or more matches to the spam list is more likely to be junk than a message with no matches. A Bayesian spam filter scans the content of messages and renders a probability that the message is spam. When the probability exceeds a set threshold, the message is marked as spam. It's not perfect, but if you check your "junk" folder, you'll see it does better than you may know.

A History of Grim Reckoning

So I picked up the *New York Times* one morning," said J. Richard Gott, "and opened to a story that says the Parthenon had been destroyed by an earthquake. I said to myself, The Parthenon has stood for thousands of years, and here I'm only twenty. What's the chance of that happening in my lifetime?"

Gott, then a Harvard undergraduate, decided the chance was very small. He was right. Pranksters from the *Harvard Lampoon,* the campus humor magazine, had printed a hoax front page with the *Times* masthead. They'd swapped it for the front page in campus subscribers' copies.

It was that mental connection, between his own span of time and that of a Greek monument, that primed Gott for his epiphany in Berlin.

Selection Effects

Gott was not the only one thinking along these lines. In September 1973 Kraków hosted a symposium in honor of the five hundredth anniversary of the birth of Copernicus. The astronomer's reputation was riding higher than ever. Copernicus is not just the guy who told us that the Earth moves around the sun. The Copernican principle, holding that our vantage point is not special, has kept Copernicus relevant in a way that even Brahe and Kepler aren't.

Like many a founding father, Copernicus is a creature of the modern imagination. Copernicus never articulated a Copernican principle, nor might such a thing have made much sense in his time. He was just trying to figure out how the solar system worked. Only in the mid-twentieth century did it become common to draw an explicit analogy between Copernicus's heliocentric solar system and later astronomical assumptions of an uncentered universe. Physicist Hermann Bondi used the term "Copernican Cosmological Principle" in a 1952 book. By the time of Gott's 1969 visit to the Berlin Wall, it was natural (for an astrophysicist) to attach Copernicus's name to a method with only a metaphoric connection to the Polish astronomer.

One of those speaking at Kraków came to bury, not praise, Copernicus-as-metaphor. Australian-born thirty-one-year-old Brandon Carter was a lecturer at Cambridge. Much of his work had explored the physics of black holes, a topic that had only lately become respectable. Carter felt that the Copernican metaphor had been taken too literally. As he put it, "Copernicus taught us the very sound lesson that we must not assume gratuitously that we occupy a privileged *central* position in the Universe. Unfortunately there has been a strong (not always subconscious) tendency to extend this to a most questionable dogma to the effect that our situation cannot be privileged in any sense."

Carter made the modest proposal that, sometimes, we *are* special. Given that we are observers of the world around us, our situation has to be special in any way necessary to permit the existence of observers like us.

This is an example of an *observation selection effect*. It's natural to assume that people, objects, and events that we can observe are typical of those we can't. That is the premise of opinion polls, which hold that a handful of randomly selected people can speak for a whole nation. But there are many ways for polls to be skewed and for observations to be distorted by selection effects.

British physicist Arthur Eddington gave a classic example in his 1939 book, *The Philosophy of Physical Science*. Wanting to know the size of the smallest fish in a pond, you get a net and scoop up a hundred random fish, measuring each carefully. The smallest fish of the hundred is six inches long.

It is easy to jump to the conclusion that fish smaller than six inches are rare or nonexistent. Nope, runs Eddington's punch line. It turns out the net can collect fish six inches or longer only. All smaller fish slip through the mesh.

"Whenever one wishes to draw general conclusions from observations restricted to a small sample," wrote Carter, "it is essential to know whether the sample should be considered to be biased, and if so how." Carter proposed that our very existence as intelligent observers imposes a bias—a net, in Eddington's analogy—restricting where we might be in space and time.

We should therefore not be too quick to assume that the Earth is a typical planet. It is only on planets where intelligent life has arisen that such matters can be discussed!

Physicists have often remarked that some of the observed universe's attributes seem improbably tailored to the origin and evolution of intelligent life. This too might be understood as a selection effect.

Carter called this thesis the *anthropic principle*. He intended the anthropic principle as a commonsense counterweight to the Copernican principle. Carter's idea has since become one of the more polarizing concepts of modern physics. It is probably fair to say that most physicists consider the anthropic principle valid but not necessarily useful. Some roll their eyes at what they see as a catchy banality that gets more than its share of media attention. "Anthropic notions flourish in the compost of lax language and beguiled thought," wrote one unsympathetic reviewer. Physicists have been hissed for using "the *A* word" in speeches. You either love the anthropic principle or hate it; you find it "deep" or a facile witticism. Wherever you go, there you are.

The anthropic principle's mixed reputation owes something to the fact that people have interpreted it in so many different ways. Carter himself offered two versions. His *weak anthropic principle* is, despite the "feeble" billing, the more important one. This is a simple selection effect saying that, as observers, we have to find ourselves in a part of the universe compatible with observers.

Carter also offered a *strong anthropic principle* saying that, as observers, we have to find ourselves in a universe whose laws permit the existence of observers. Though also a truism, it borders on the metaphysical, and Carter wrote that it is "not something that I would be prepared to defend with the same conviction" as the weak version.

Many further iterations of Carter's idea have been proposed. Two of the most ardent proponents, John Barrow and Frank Tipler, devised the Final Anthropic Principle (FAP): "Intelligent information-processing must come into existence in the universe, and, once it comes into existence, it will never die out."

Writing in the *New York Review of Books,* Martin Gardner quipped that FAP might better be called the Completely Ridiculous Anthropic Principle (CRAP).

By 1983 Carter had found another application of anthropic-style reasoning: predicting the future survival of the human race. In a 1983 lecture at the Royal Academy, London, Carter described what we now call the doomsday argument. He believed it to be "an application of the anthropic principle outstandingly free of the questionable technical assumptions involved in other applications" and "obviously the most practically important application."

Yet Carter did not fully accept the grim forecast of his math. Nor did many others. While the anthropic principle drew spirited debate, the doomsday thesis was roundly rejected. Indeed, the doomsday discussion is omitted from the printed version of Carter's lecture. Carter chose not to publish on doomsday, discussing it only in seminars where he thought it could get a fair hearing. In this way the doomsday argument began as a secret, almost samizdat doctrine, known to a few as the "Carter catastrophe."

Why Is There Anything?

Fresh out of Oxford, John Leslie took a job as an advertising copywriter at McCann-Erickson's London office. He stuck with it long enough to realize that he wanted to think deeper thoughts than the advertising business needed. Quitting his job, he studied philosophy at the University of Guelph, Ontario, Canada.

Leslie was an active outdoorsman—a rock climber, canoeist, and explorer of volcanoes—and an enthusiast of the games of Go and chess. He created a board game called Worldmaster, marketed in 1989, that was something like a cross between Risk and Scrabble. Players conquer countries by spelling their names with letter tiles. Leslie also invented Hostage Chess, a much-studied variant in which captured pieces are hostages that may be exchanged and returned to the board.

Now retired from teaching, Leslie lives with his wife in a verdant part of Victoria, British Columbia. He speaks with a clipped British accent and has a habitually puckish expression. The focal point of Leslie's long career has been the biggest question of all: Why is there anything (a universe rather than nothingness)? Science journalist Jim Holt rated Leslie the "world's greatest expert" on this almost indefinable topic. Yet Leslie is no less known as an expert on when the world might end. His interest in that topic came about through a September 1987 meeting with physicist Frank J. Tipler.

The Vatican had organized a meeting of scientists and theologians at Castel Gandolfo to mark the three hundredth anniversary of Newton's *Principia*. "Tipler was one of my special pals there," Leslie recalls. Alabama-born, Tipler was trained at MIT and the University of Maryland. Tipler's achievements in cosmology have long since been overshadowed by his enthusiasm for wild ideas. He is best known for his "Omega Point" hypothesis, which says that our exponentially growing computing power will eventually lead to an omniscient, omnipotent, and omnipresent singularity, attaining the traditional attributes of God.

As his many critics see it, Tipler is a specimen of that rare breed, the tenured crank. As a professor at Tulane University he teaches a class in Omega Point Theory (PHYS 1190) as well as introductory physics. Tipler has expressed doubts about the evidence for Darwinism and global warming. When Michael Shermer came to write a book titled *Why People Believe Weird Things*, he devoted an entire chapter to Tipler.

Carter's doomsday argument set off Tipler's weird-idea radar. And given Carter's reservations, Tipler may have been the first to fully accept the doomsday argument. Tipler is earnest and animated, with something of the quality of a door-to-door salesman who truly believes in the product. At the Rome meeting, he described the doomsday argument for Leslie. The philosopher "became convinced of its importance after an initial two minutes of thinking that it just had to be wrong." With that, Leslie too was converted to the small core of doomsday believers.

When Leslie began a correspondence with Carter, the physicist made an unusual request. He asked that Leslie refer to the idea as the "Carter-Leslie doomsday argument" to share "not only the credit but the blame, which will not be in short supply."

Ostatni Dizien

Carter encouraged Leslie to publish the doomsday argument, saying that he would be "muttering support from the trenches." Doomsday made it into print twice in May 1989. Leslie published a short description of Carter's idea in "Risking the World's End" in the *Bulletin of the Canadian Nuclear Society.* The same month Danish physicist Holger Bech Nielsen, a pioneer of string theory, outlined the idea in a physics paper, "Random Dynamics and Relations Between the Number of Fermion Generations and the Fine Structure Constants," in the Polish journal *Acta Physica Polonica.* The article records four lectures Nielsen gave the previous year in Zakopane, Poland. The doomsday discussion is the second half of the third lecture.

Nielsen wrote in English (and math), using the word "doomsday" and supplying its Polish equivalent: "Now my point is that this procedure leads us to discard all sensible scenarios with the exception of those which either have a violent end, Doomsday (*Ostatni Dizien*), or at least such a strong reduction of population that it would never again rise to the present height, i.e., also a Doomsday. Estimates show that this 'Doomsday' must...come not later than a few hundred years from now."

Ostatni Dizien is Polish for "Last Day." Nielsen offered the first thunderclap of the brewing storm: "It is a pleasure to thank N. Brene for producing these notes from my draft," he wrote. "He does not, however, feel responsible for the contents of the third lecture."

Despite the fact that Nielsen is a celebrity scientist in Denmark, the article did not get much attention for the doomsday argument. The discussion was buried in a highly technical paper. Nationality and culture matter, even in the ever-globalizing world of science. The 1989 discussions in Canadian and Polish journals probably limited their audiences.

Leslie went on to publish doomsday articles in the *Philosophical Quarterly* (1990) and *Mind* (1992). Gott's *Nature* article, "Implications of the Copernican Principle for Our Future Prospects," appeared in 1993. Scientists around the world read *Nature,* as do science-minded journalists looking for feature stories. With these high-profile publications, the doomsday conversation began in earnest.

"Implications of the Copernican Principle"

In the summer of 1990 Gott called his college friend Chuck Allen. "Chuck, you remember that prediction that I made about the Berlin Wall? Well turn on your television!" NBC anchor Tom Brokaw was reporting live from Berlin. The wall was coming down.

"I thought, well, you know, maybe I should write this up," Gott said.

He was not so much concerned about staking out priority (he was unaware of Leslie's and Nielsen's publications). Instead, it worried Gott that the history of science holds many instances in which someone never got around to publishing or promoting an idea that died with them, taking years or centuries to get its due. Bayes's theorem is one example. The one that was on Gott's mind was Hero's engine. In the first century AD Hero of Alexandria described a simple steam engine. It wasn't until seventeen hundred years later that a similar idea was put to widespread practical use. It was, furthermore, the nuts-and-bolts engineering of steam engines that motivated the science of thermodynamics.

Gott wrote a paper on his delta *t* argument and ambitiously submitted it to *Nature,* whose editors sent it to referees, one of whom was Brandon Carter. Through Carter, Gott learned of Leslie's and Nielsen's publications. But Gott developed the idea in several new directions. In its six pages Gott's article treats not only the future of the human race but space travel and the search for extraterrestrial life. Gott begins by saying that

> the location of your birth in space and time in the Universe is privileged (or special) only to the extent implied by the fact that you are an intelligent observer, that your location among intelligent observers is not special but rather picked at random. Knowing only that you are an intelligent observer, you should consider yourself picked at random from the set of all intelligent observers (past, present, and future) any one of whom you could have been.

This is a statement of what is now called the *self-sampling assumption*, or the *human randomness assumption*. Gott used it to estimate the longevity of our species. "Disturbingly, even extraordinarily low values [for future time until doomsday] cannot be confidently excluded," Gott wrote, "but high values..., such as many billion years, which we might hope for, can be" ruled out.

"The methods that I have used here are very conservative; if the results are dramatic it is only because the facts are dramatic.... This paper only points out and defends the hypothesis that you are a random intelligent observer....Short of having actual data on the longevities of other intelligent species, this hypothesis is arguably the best we can make."

Copernicus challenged the Church teaching that the Earth is the center of God's creation. Gott challenged one of the core doctrines of secular, technocratic society: that humanity has a long future that will likely entail space travel and settling on many other planets. Gott estimated the probability that we will settle on other planets throughout the galaxy at a mere 1 in a billion.

Copernicus and Galileo defied the Inquisition; Gott took on the gospel of *Star Trek*.

"'There are lies, damn lies and statistics' is one of those colourful phrases that bedevil poor workaday statisticians," complained Steven N. Goodman, biostatistician at Johns Hopkins. "In my view, the statistical methodology of Gott... breathes unfortunate new life into the saying." That was one of several critical, even exasperated, letters to the editors of *Nature*.

The discussion spilled over into the general media. There was a favorable feature on Gott and his ideas in the *New York Times*. The following month the *Times* published an op-ed sharply critical of the doomsday argument. Writer Eric J. Lerner, a gadfly physicist who disputes the big bang, branded Gott's *Nature* article "pseudo-science, a mere manipulation of numbers to disguise an implausible argument. Why would a prestigious journal like *Nature* publish such astrology and why would a prominent cosmologist, who presumably knows better, write it?"

Lerner answered his own question. "History shows that whenever a society stops advancing, when the standard of living falls, as is the case today, there are always so-called experts who rush forward to absolve the powers that be of any responsibility... [for] the greed and shortsightedness of those who rule."

Lerner was a socially engaged activist who had marched for civil rights in Selma. He painted Gott as a kook, but beyond that, many *Times* readers must have been mystified as to what Gott was saying (astrology for capitalist overlords?).

Gott's tart response, published as a letter to the editor, noted that

> Mr. Lerner refuses to believe that he may be randomly located among human beings.... This is surprising since my paper had made a number of predictions that, when applied to him,

> all turned out to be correct, namely that it was likely that he was (1) in the middle 95 percent of the phone book; (2) not born on Jan. 1; (3) born in a country with a population larger than 6.3 million; (4) not born among the last 2.5 percent of all human beings who will ever live (this is true because of the number of people already born since his birth)....Mr. Lerner may be more random than he thinks.

In 1996 John Leslie published *The End of the World: The Science and Ethics of Human Extinction*. It was the first book to present the doomsday argument at length, alongside a dizzying catalog of potential disasters, from the familiar to the exotic. Where Gott was a stoic, Leslie saw the math as a wake-up call. We have the power to change the prior probabilities of extinction, he maintained, and the ethical obligation to try to do so. As Ebenezer Scrooge said to the Spirit of Christmas Future, the doomsday argument is a shadow of things that may be, not must be.

Leslie's book received a scorched-earth review in *Nature,* the journal that had launched the doomsday controversy. The reviewer was the renowned physicist, mathematician, and author Freeman J. Dyson, of the Institute for Advanced Study, Princeton. "After careful consideration, I state unequivocally that the application of the Bayes rule here is invalid," Dyson wrote. "This discussion is worthless."

Dyson went on to compare Leslie's book to Malthus's famous *Essay on the Principle of Population* (1798). He did not mean that as a compliment. "Uncritical belief in Malthus's predictions helped to hold back political and social progress in Britain for a century," Dyson said. "Because of this unhappy precedent I consider it important to call attention to the fallacy in Leslie's argument."

Leslie defended his book in a letter to *Nature* and, several years later, Carter came to its defense as well. Dyson was "apparently under the influence of wishful thinking," Carter wrote. "I have found however that such conclusions tend to be unpopular in many quarters, presumably because they involve limitations on the extent and more particularly the duration of civilizations such as ours which (in lieu of personal immortality) many people would prefer to think of as everlasting."

Here Carter was alluding to Dyson's concept of "eternal intelligence." In a speculative 1979 paper, "Time Without End: Physics and Biology in an Open Universe," Dyson outlined a way in which intelligent life might conceivably evade entropy and survive forever, past the last flickering of stars and the heat death of the universe. Technologically adept observers might be able to reengineer themselves so that they could experience a subjective eternity, even as the universe cooled off to absolute zero. The result would be "a universe growing without limit in richness and complexity, a universe of life surviving forever."

There are many reasons to question the feasibility of Dyson's idea, but the doomsday argument added a novel one: if human consciousness is to survive for quadrillions of years, then it's weird that we're on page one of that glorious multivolume history. Carter's insinuation was that Dyson was "subconsciously" taking his pet ideas for granted and not giving Leslie's (and Carter's) ideas a fair reading.

It is unusual for those originating an influential idea to not assert credit for it. Carter has been curiously reticent about the doomsday argument. He sardonically spoke of being assassinated, so unpopular was the idea. Only in recent years has he alluded to the doomsday argument in discussions of the anthropic principle. Here, in his own words, is his concise description of the doomsday argument, from a 2004 talk in Paris: "The anthropic principle's attribution of comparable a priori weighting to comparable individuals within our own civilization makes it unlikely that we are untypical in the sense of having been born at an exceptionally early stage in its history, and hence unlikely that our civilization will contain a much larger number of people born in the future."

Carter describes this as "a thesis developed particularly by Leslie (and from a slightly different point of view by Gott)," leaving his own role unmentioned.

The Postapocalyptic Future

Should World War III start tomorrow, it probably would not kill everyone. ("I'm not saying we wouldn't get our hair mussed," as the general says in *Dr. Strangelove.*) But a major nuclear war and its aftermath would disrupt agriculture, trade, and infrastructure, ending civilization as we know it.

Willard Wells, a Carlsbad, California, physicist, asserts that the standard doomsday argument places too much emphasis on human extinction. More likely the future will be postapocalyptic. In his 2009 book *Apocalypse When?* Wells applies Copernican reasoning to civilization. Our urban society, he says, can trace its roots back to the one that arose in Mesopotamia about 11,000 years ago. Most civilized people have lived in the past few centuries. Being much younger than *Homo sapiens,* our civilization is likely to have a shorter future, all else being equal. Wells estimates the median future duration of civilization to be about 860 billion person-years. (That's the total of years yet to be lived under civilization, for all the world's people.) At today's population, Wells's estimate corresponds to only about 115 more years of civilization. Wells therefore believes that the chance of civilization ending is somewhere around 1 percent per year.

It is not easy for our planet to support billions of people. This is the juggling act of a finely tuned global economy that moves food and goods across continents and oceans. If anything were to happen to that global economy, billions could die of starvation. The small postapocalyptic population would then slow down the hands of the doomsday clock. Human extinction might be put off a long time, but billions would have died.

Wells's estimated 1 percent per year chance of societal collapse is greater than the chance that an average home will burn down this year. We take that seriously enough to buy insurance. Parents worry about unsafe car seats and vaccination side effects and doctored Halloween candy. Wells says there is more cause to worry that a child born to an affluent first-world family today will starve to death in a postapocalyptic hellscape. The survivors would find that affluent world they were born to, of endless cable channels and hipster food trucks, gone forever. Wells's conclusion is unsparing: "And so the short answer to the big question is, No. There is no way out of our dilemma. An apocalyptic event, perhaps a near-extinction, is prerequisite for long-term human survival, and that's just the way it is."

I will close this mini-history by mentioning two more people who seem to have independently conceived the doomsday argument: Stephen Barr, an American particle physicist also known for his writings on science and religion, and Saar Wilf, an Israeli tech entrepreneur and competitive poker player. There are many cases of simultaneous discovery in science: Newton and Leibniz (calculus), Le Verrier and Adams (Neptune), and Darwin and Wallace (evolution). Those famous examples involve two people coming up with nearly the same idea at nearly the same moment. The doomsday argument may have no fewer than five codiscoverers. In the closing decades of the twentieth century, eschatology was in the air.

Twelve Reasons Why the Doomsday Argument Is Wrong

Is the Doomsday Argument correct?" asked Dutch physicist Dennis Dieks. "Nobody I told about the Argument was prepared to think so. But no one was able to offer a clear and convincing view about the nature of the error (if there is one.)"

"I have encountered over a hundred objections against the doomsday argument," wrote Nick Bostrom, "...many of them mutually inconsistent. It is as if the doomsday argument is so counterintuitive (or threatening?) that people reckon that every criticism must be valid."

"Given twenty seconds, many people believe they have found crushing objections," wrote John Leslie. "At least a dozen times, I too dreamed up what seemed a crushing refutation of it. Be suspicious of such refutations, no matter how proud you may be of them!"

These reactions are typical. Upon hearing of the doomsday argument, most of us think it's obviously wrong, and it's easy to spot why it's wrong. But it's not so readily refuted as it seems. This has made doomsday a perfect fit for philosophy journals. "One odd thing is that you very seldom get papers saying somebody is right," Leslie told me. "It's so much easier to get a paper published saying someone is wrong." The stream of doomsday papers continues to this day.

A good way to get up to speed on the debate is to run through the commonest refutations (and explain why they may not slay the beast).

I'm Not Random

The "big question," as Leslie wrote, is "whether we have any right to treat *being born at a particular time in human history* as at all analogous to *having one's name come out of an urn*." This part of the doomsday premise leaves many uneasy.

Random drawings are expected to be well mixed. Should I be asked to draw a lottery number from an urn, on TV in view of millions of ticket holders, I wouldn't just pick the ball that happened to be right on top. I'd make a big show of mixing up the balls, bringing the ones on the bottom to the top and vice versa. I'd perform the randomness.

Past-present-future human lives can't be mixed up that way. Each of us has a unique identity that is bound to the age in which we live. I'm not a frontier homemaker of 1850s Dakota Territory or a wormhole technician of the thirty-seventh century. Had I been such a person, living in a completely different culture, I'd be someone else, not me. We are all pinned like butterflies to our place in history.

"I'm not random" is another way of saying that there can be no randomness without a randomizing procedure. This sounds like a reasonable claim. Yet one person's *random* is another's *methodical*. Take Gott's phone book example. It's a safe bet that an arbitrary name falls in the middle 95 percent of the book. My name does. This is incontestably true, and randomness has nothing to do with it. The names are in alphabetic order! Furthermore, I am who

I am. I'm not AAA Pest & Termite or Theodore R. Zyskowski. It is not necessary to pretend that I could have had a name falling somewhere else in the alphabet. Gott's claim works nonetheless.

Self-sampling helps explain some of life's minor mysteries. Why is the next lane of traffic (next line at the bank, the market, the motor vehicles department) "always" faster? Studies have shown these perceptions are real, not just psychological.

The explanation is simple enough. Lanes and queues are slow because they're crowded. It's not always easy or safe to switch lanes, so drivers tend to remain in crowded lanes awhile. (At the bank, switching means moving to the back of the new line. Nobody likes that.) So, if you are an arbitrary motorist out of all the motorists on all the highways of the world, you are more likely to be in a lane with more cars than fewer cars. Meaning a slow lane.

Could you, in your electric blue 2004 Mini Cooper with license plate ANZ 912, possibly be someone else? No. You're unique. But if you want to understand why the adjacent lane is faster, it may be helpful to consider yourself a random draw.

Now Isn't a Random Time

The doomsday argument also depends on the claim that *now* is a random point in time. William Eckhardt, a mathematically trained commodity trader who has written on doomsday, points out that ideas and inventions do not arise at any random point in history. They reflect the needs of the cultures that create them. We shouldn't be surprised that the doomsday argument was devised shortly after the appearance of nuclear weapons, genetic engineering, global climate change, and the first efforts toward artificial intelligence. Life has always been fragile, but such developments have put us in an unsettling new place.

That Gott, Carter, and Nielsen had similar ideas, independently, about the same time, is no coincidence. It's a zeitgeist. The present epoch can't be regarded as random, so far as human extinction is concerned.

We can deconstruct the doomsday argument as a cultural artifact, an expression of our age's anxiety about the future. This undercuts the doomsday argument as we've laid it out, but it may not get us off the hook. If our topical concerns have any rational basis at all, then the takeaway may be similar: the end is near.

What About Adam and Eve?

Adam and Eve (or a Cro-Magnon, etc.) might have applied the doomsday reasoning to predict extinction occurring before the twenty-first century. Yet here we are, alive and kicking. Doesn't that prove there is something wrong with the doomsday argument?

When asked *What about Adam and Eve?* J. Richard Gott goes full Socratic. "Are *you* Adam or Eve?" he asks. His rebutters never are. And that's the point. Some people have to be very, very early in the birth-order lottery. But it's unlikely for *me* to be very, very early.

The doomsday argument is a statement of probability. It's like a weather forecast saying there's a 70 percent chance of rain. If it doesn't rain, you may say the forecast was wrong. That's not quite fair. The forecast allowed for a 30 percent chance of no rain. Statements of probability ought to be judged by how well calibrated they are. You can test weather forecasts to see how accurate their claimed percentages are in the long run. We can't test predictions of a one-time event like human extinction.

The Copernican method itself can be tested. I'll get to that in a coming chapter.

Somebody Has to Be "Early." Why Not Me?

The staff of a start-up is working on an app that they hope will have a billion users one day. I'm a beta tester, the seventh person to get the app. A strict Copernican might say that I can be highly confident there will never be more than a few hundred users. A venture capitalist who accepted that logic would never invest in any app!

This is a variation on Gott's wedding crasher. As a beta tester, I am not a random user. I know that I am an early adopter.

There may be those who feel they can intuit the basic plot of human existence. They believe we are still early in the story arc. To the extent that anyone can be *sure* of this, the doomsday argument is irrelevant.

The catch is that it takes incredible confidence in our being early (typically, over 99 percent) to downsize the doomsday risk significantly. Is anyone *that* sure of our future?

There Is No Master List of Humans

The Carter-Leslie doomsday argument asks us to imagine a complete list of all past, present, and future humans. This list is a fiction. It does not exist. It may never exist.

I know I'm not Captain James Tiberius Kirk of the space-faring future. I can't even name *any* specific, nonfictional person from a century yet to be. Why then should I reason as if the names of all the future people could be put in a hat, along with my own and all the names of the past people, for a random drawing?

Leslie offered this counterargument. Imagine that a secret, well-funded foundation sets out to award 5,003 emeralds to 5,003 randomly selected winners. Three of the winners are to be in one century, and the other five thousand in a later century. The names of the winners are never announced, and each takes a vow of silence.

You're an emerald winner. You don't know whether you're in the earlier century or the later one. Since there are only three winners in the earlier century and five thousand in the later one, the odds strongly favor you being in the later century.

If you were offered an even-money wager, you should bet you're in the later century. If all 5,003 people did that, five thousand would win the wager and only three would lose. This beats betting the other way.

In the earlier century, the emerald foundation's list would have three names plus five thousand blanks to be filled in later. Only in the later century could the list be complete. But that shouldn't affect the reasoning of the emerald winners. The key point is that I don't know anything about my position on the list. I resort to self-sampling because of my ignorance, not in spite of it.

Self-sampling does not require a roll of the dice or a spin of the chuck-a-luck. It's not necessary to picture myself as a disembodied soul plunked down into a random body in a random century. Least of all do I need to fret that I "could have" lived in a different time. The emerald winners know who they are and what year it is. They just don't know where they fit into the *relative* chronology of the emerald experiment. It is ignorance about one's contextual situation—not the ability to hopscotch freely across centuries—that justifies self-sampling.

Doomsayers Make Inconsistent Predictions

A good reason why thoughtful people don't believe in astrology is that it fails to make consistent predictions. One horoscope says it's a good day for a Libra to start a business venture. Another says the opposite. They can't both be right. The doomsday argument can raise similar qualms. If I assume I'm at a random point in the duration of the human race, I get one prediction. If I say I have a random birth rank, I get a different one. Factor in my prior probabilities, and there's still another. Which should I believe?

All three versions of the doomsday argument share the assumption that I am not improbably special. There are many ways to be unspecial. I probably like pizza; I probably hate the sound of fingernails on a chalkboard; I probably didn't front an emo band in the 1990s. . . . All are safe bets. This list of safe bets could be continued indefinitely. But if we were to add enough such usually true claims to the list, I (anyone) would turn out to be improbably "special" in *some* ways.

Both the regular-clock and birth-clock versions of doomsday say that my time of birth was probably not special. Most likely both versions yield correct predictions about human extinction, though it's possible that just one does, or even that neither does.

One thing the doomsday arguments demonstrate is that having an average birth rank can make me atypically late, by a clock of regular time. The chart shows how. As before, I chart time on the horizontal axis. But now the height of the curve is population, and the area under the curve represents the cumulative population. The "average" human is not in the middle, at time 50, but later.

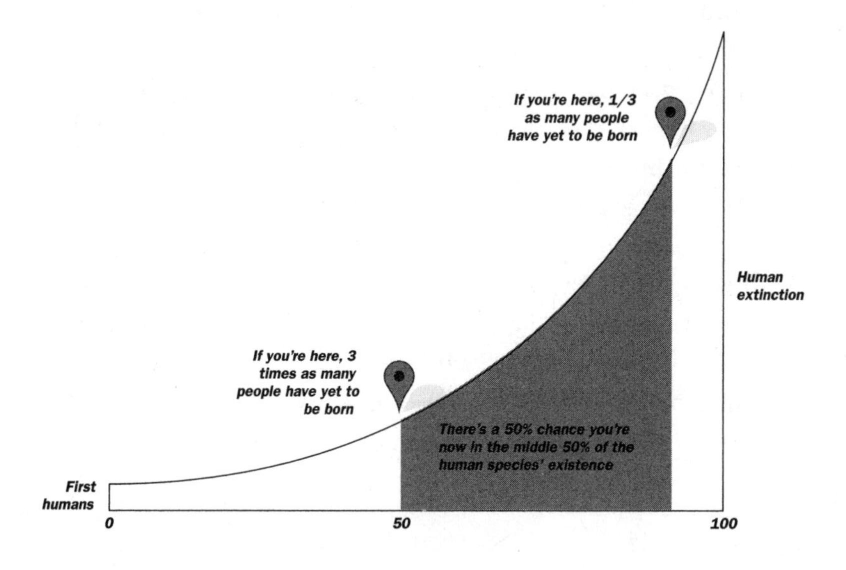

The shaded region is the middle 50 percent of the curve's area. Because most observers skew to the right, extinction is closer than the species' origin is—for most observers. Humanity is slouching toward Armageddon.

This population curve is just one possible model. You can draw any curve you want, so long as the leftmost part is consistent with the population explosion of recent times.

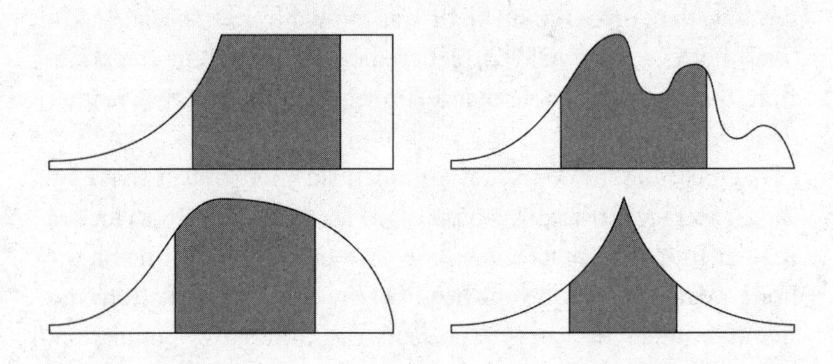

Hence a catch-22 of birth-clock doomsday calculations: in order to get a date for doomsday, we must supply *future* demographic data. We need years and centuries of future census figures. But if we had that, we wouldn't need the doomsday argument. We'd just check those future census records to see what year the population hits zero.

Yet even this should not be regarded as disqualifying. What the doomsday argument really says is that I am unlikely to be extraordinarily early, by births *or* time. From that it follows that there is a trade-off. The future could be populous (but only if doomsday is imminent). The future could be long (but only if its population is small). Either way, humanity is highly unlikely to ever attain a cumulative population of 10 trillion. Right or wrong, this claim is not vacuous.

The Doomsday Argument Is Unfalsifiable

Critics such as Eric J. Lerner have complained that the doomsday argument is *unfalsifiable.* That buzzword, coined by philosopher Karl Popper, reflects the view that a legitimate scientific theory must be one that can potentially be proven wrong. "Unfalsifiable" has become one of the most damning things a scientist can say about another's theory. For that reason, it's said a lot.

You may think Popper had it mixed up. Isn't science about proving theories *right*? No, said Popper. Theories are generalizations. Generalizations can never be proven right. Should you travel the world looking at ravens, and they're all black, this would give you confidence in the rule "All ravens are black." But it wouldn't prove it. There could be orange ravens somewhere you didn't look. Proving a generalization is like Zeno's paradox: the runner eternally falls short of the goal. For Popper absolute truth is a mirage we never reach.

It is rather falsifications that are decisive. Find one orange raven, and "All ravens are black" is wrong.

There are good reasons for scientists to favor readily falsifiable hypotheses. It's too easy to spend one's career on dead ends as it is. But in the real world, almost no one proposes a hypothesis that's *easy* to falsify. Were it possible to prove a claim wrong with a $10 experiment, the proposer would have already done that experiment. The theories that get published tend to require a lot of time, ingenuity, and resources to test. There are never enough resources, and the theorists have the job of lobbying experimentalists, persuading them that their ideas are interesting enough to test. Sometimes that means deprecating rival theories as "unfalsifiable."

"Like any good scientific hypothesis, this hypothesis is falsifiable," Gott wrote of the doomsday argument. It could be disproved "if more than 2.7×10^{12} more human beings are born." It's more accurate to say that doomsday is unverifiable. We can't be certain the prediction is right until, uh, everyone is dead.

You Can't Predict from Old Evidence

A Bayesian uses new evidence to update prior probabilities. With doomsday, the new evidence isn't all that new. I knew what century it was before I ever heard of the doomsday argument.

This is known as the *old evidence problem*. We are reluctant to make inferences from evidence that has long been known. I can't even remember when I learned what century it was. I must have been exposed to the concept of centuries several times in childhood before I caught on to what the grown-ups were talking about.

Even for humanity in the aggregate, it was a slow, gradual slog to establish our species' position in the geological and cosmic time scales. Cumulative population figures have been refined over many years. There was never a single moment of truth in which the doomsday evidence was revealed.

A little thought shows this concern to be irrational. Just today I was dusting that kickball trophy I won in third grade (p. 37). I got to thinking: that's funny because I don't recall being any good at kickball.... Then it hit me: it was a participation trophy!

This may well be a valid Bayesian conclusion. Its validity stands regardless of how long it took me to make the mental leap. Better late than never!

Leslie put it this way: "Dear Mr. Newton: What's all this nonsense about deriving a new physics from the fall of an apple? Surely you've long known that apples do fall."

You Can't Predict from Subjective Evidence

The "you are here" pin exists on a different plane from the objective information of the underlying map. Self-locating information is subjective and first-person. That puts it at a remove from the realm of science and objective fact. Even the grammar of scientific papers enforces this distinction. Journal articles are to be written in a neutral third person ("The investigator demonstrated..."). There is no "I" in science's team.

Yet self-locating evidence is as real as any other kind. Clocks, calendars, road signs, compasses, and GPS were invented to supply self-locating information. *Of course* that information is useful, with real-world implications. When I want to know whether I'll be late for a meeting, I check my clock or Google Maps. We use self-locating information for predictions all the time.

A Society of Immortals Could Cheat Doomsday

A technological optimist can imagine that medical science will cure all diseases and aging in the next century or two (time we've probably got, before the end...). A society of virtual immortals might not need to have children. The birth rate could plummet. We might never arrive at doomsday's final birth number, *Z*. Saved by the bell?

Unfortunately, no. The doomsday argument need not be framed as a prediction about births. It is most fundamentally about moments of consciousness ("observer-moments"). Sometimes this is a hairsplitting distinction we can ignore. As long as lifespans are finite and don't vary too much, births are a handy measure of human lives and experience. But for an immortal society with no births, we would have to reconsider what it means to be a random sample. It would make sense to regard my present moment of consciousness as a random draw from the whole collective stream of consciousness of the human race. The doomsday argument then says that my moment is unlikely to be extraordinarily early, and this sets approximate limits on how many conscious moments there are likely to be in the future.

Observer-moments are much like the economic concept of worker-hours. Multiply the number of minds by the number of hours that those minds exist and are conscious. An "immortal" who lived seventy thousand years would have a thousand times the observer-moments—meaning, a thousand times the thoughts and experiences—of a mortal who lived seventy years.

It follows that a stable population of 10 billion immortals would have the same number of observer-moments as a continually replaced population of 10 billion mortals. The doomsday math would not change. Even for would-be immortals, the end is near.

We'll Evolve into Something Better

The dodo birds of Mauritius were first reported in 1598. Within about seventy years, European sailors had killed them off. The last sighting of a live dodo was in 1662.

Not all species end like the dodo, in an abrupt extinction. *Homo heidelbergensis* gradually evolved into *Homo sapiens*. From our perspective, that was a good thing. We flatter ourselves that we have more meaningful inner and outer lives than *Homo heidelbergensis* did, with better food, better clothes, better entertainment, better everything that matters. This advancement may continue. There may come a time when our descendants are different enough from us to rate their own, more fabulous species.

This is a hopeful thought. But biological evolution is slow. There's not much prospect of evolving into a new species in the time frame of the population-based doomsday argument.

More believable is that humans might adapt with the aid of technology. Genetic engineering, robotics, and artificial intelligence could transform humanity into something radically different in the coming centuries or millennia. Biological humans might be verging on obsolescence, tech boosters say, but that's okay—so long as human-like consciousness lives on in other, better vessels ("posthumans").

This raises the question of how technologically enhanced beings would fit into the doomsday argument. I will come back to that issue later. For the moment let's say that it doesn't appear to offer an easy out. The doomsday argument does not depend on any narrowly legalistic definition of the word "human." We can go back and replace every mention of "human" with "human or posthuman." We would then come up with the same numbers of past minds or observer-moments, leading to nearly the same result. It's not just us but our technological heirs and assigns who fall under the shadow of doomsday.

We Should Go Out and Get More Evidence

"In the absence of data, we are told to follow Gott's [method]," wrote philosopher of science Elliott Sober. "I'd expect most biologists to say something different—in the absence of data, you should go out and get some."

A physicist colleague of Leslie's struck a similar note, objecting that "you can't get any result from a single trial."

Leslie responded with a thought experiment. A certain interesting atom is produced at the cost of $1 billion. Theory says the atom is likely to decay (break up into other particles) in about a second *or* about a hundred billion years. The atom decays in a second. "Will you repeat the experiment at the cost of another billion dollars?"

The physicist reiterated, *You can't get any result from a single trial*—"at which point," said Leslie, "I just gave up."

It is no coincidence that many of the doomsayers are astronomers and cosmologists. A biologist can run off as many bacterial colonies as desired. Particle physicists just need money to do great things. But an astronomer is stuck with the awkward circumstance that there's only one universe.

Much as the Plains Indians made use of every part of the buffalo, astronomers and cosmologists prize every scrap of data they've got. That includes self-locating information and its Bayesian implications.

No one likes to draw conclusions from a single event. No scientist should do so when the event is easily repeatable. Not everything in science or life is. If we want to know how long the human race will last, we *should* go out and get more data about the survival rates of the universe's intelligent species. This, however, is easier said than done.

Twenty-Four Dogs in Albuquerque

Carleton Caves returned from a sabbatical to discover *The New Yorker*'s 1999 piece on J. Richard Gott III in his mail. He found it "incredibly irresponsible" of the magazine to promote Gott's thesis. "Anybody can see it's garbage."

So on October 21, 1999, Caves, of the Center for Advanced Studies, University of New Mexico, sent out an email to faculty, staff, and grad students. He was looking for dogs—*old* dogs—to settle a scientific question. The dogs would not be harmed. Caves was a quantum physicist.

"Gott dismisses the entire process of assembling and organizing information about a phenomenon," Caves wrote. "Put succinctly, he rejects as irrelevant the process of rational, scientific inquiry, replacing it with a single, universal statistical rule." He concluded that "it was important to find the flaws in Gott's reasoning: flawed thinking is an inevitable, even necessary part of the scientific enterprise, but when it makes its way into *The New Yorker,* the time has come to find the flaws and draw attention to them."

To that end, Caves compiled "a notarized list of...24 dogs,

including each dog's name, date of birth, and breed, and the caretaker's name." Six of the dogs were ten years old or older. Caves planned to offer Gott a $1,000 wager on each dog's life, $6,000 in all. The bet would be on when the dogs would die.

Seventeen-Year Cicada

Caves saw Gott's formula for predicting "everything" as a new brand of snake oil. He wrote:

> Gott is on record as applying his rule to himself, Christianity, the former Soviet Union, the Third Reich, the United States, Canada, world leaders, Stonehenge, the Seven Wonders of the Ancient World, the Parthenon, the Great Wall of China, Nature, The Wall Street Journal, The New York Times, the Berlin Wall, the Astronomical Society of the Pacific, the 44 Broadway and off-Broadway plays open and running on 27 May 1993, the Thatcher-Major Conservative government in the UK, Manhattan (New York City), the New York Stock Exchange, Oxford University, the Internet, Microsoft, General Motors, the human spaceflight program, and Homo sapiens.... Although Gott issues occasional cautionary statements about the applicability of his rule, the list of phenomena to which he has applied the rule indicates that these cautions don't cramp his style much.

There is, as we've seen, a distinction between Gott's delta *t* or Copernican method (which can apply to durations generally) and the Carter-Leslie doomsday argument (which uses prior probabilities of human extinction). Since the 1990s the doomsday literature has largely focused on the Carter-Leslie argument, sometimes dismissing Gott's version. Nick Bostrom offered this curt assessment: "We can distinguish two forms of [the doomsday argument] that have been presented in the literature. . . . Gott's version is incorrect."

It might be fairer to say that Gott strikes a different trade-off between simplicity and generality. The Copernican method is like the sleek tech gadget that comes with intelligent defaults. The Carter-Leslie argument promises to be more customizable, more suited to those who like to tinker.

Gott's 1993 article does not mention Bayes's theorem or prior probabilities. For some *Nature* readers that was a great sin. I asked Gott why he omitted Bayes, and he had a quick answer: "*Bayesians.*"

"I didn't put any Bayesian statistics in this paper because I didn't want to muddy the waters," he explained. "Because Bayesian people will argue about their priors, endlessly. *I* had a falsifiable hypothesis."

The long-standing complaint is that prior probabilities are subjective. A Bayesian prediction can be a case of garbage in, garbage out. There is plenty of scope to slant the results to one's liking, and to wrap them up in the flag of impartial mathematics. For Gott the Copernican method's clean, prior-free interface is a feature, not a bug.

Carleton Caves is a Bayesian if anyone is. He's a proponent of an interpretation of quantum theory known as quantum Bayesianism. Caves insists on the importance of priors. Hence the dog wager. Gott's Copernican method assigns a 50 percent chance to any randomly encountered ten-year-old process surviving another ten years. Should that process be a chocolate Lab answering to "Bella," the prediction is almost certain to be wrong.

Caves offers a compelling breakdown of the delta *t*/Copernican method. Gott folds together two distinct claims, he says. The first applies to situations in which you encounter a process at a random moment in its duration and don't know how long that process has been going on (much less how long it will continue). In that state of near-complete ignorance it's legitimate to make claims like this: the chance that I'm observing this process in the first half of its existence is 1/2. Or, the chance of my being in the first 1/*X* fraction of the duration is 1/*X*. Caves agrees with Gott on this.

In order to make predictions in ordinary time units like seconds or years, or even in less conventional units like births or movie franchise sequels, it is necessary to learn the past duration of the process. Gott's second implicit claim is that learning the past duration does not render the whole prediction superfluous.

How can that be? Consider the seventeen-year cicada—an American insect that lives seventeen years underground, then emerges for a few weeks to screech annoyingly, mate, and die. Essentially all these cicadas live seventeen years. Should I be digging a new water line and come across a seventeen-year cicada at a random point in its existence, there's a 50 percent chance it's in the first half of its lifespan. This is the first claim Gott is making, and it's unquestionably right.

Now let's say I learn my random cicada is eleven years old. I know its biological clock has exactly six more years to run. A Copernican prediction is moot. It's irrelevant not because it would be wrong (actually it would be *right*) but because it can't match the degree of confidence and exactitude I already have, knowing what I do about seventeen-year cicadas.

There aren't many creatures with such precisely defined lifespans. But the seventeen-year cicada illustrates why the Copernican method is useful in some cases and not in others. The reason, says Caves, is *scale invariance*. We need to be dealing with a process that has no characteristic time scale or lifespan, or at any rate, none that we know about.

Fractals and Scale Invariance

"Scale invariance" may be an unfamiliar term. Here's one more likely to ring a bell: "fractal." That word was coined by Benoit Mandelbrot to describe the fascinating unruliness of nature. Coastlines, snowflakes, clouds, and landscapes resist the straitjackets of Euclidean geometry. A coastline is not a "line." A snowflake is not a hexagon. The defining quality of a fractal is scale invariance, or self-similarity. When a picture or diagram or chart of a fractal is zoomed in or out, its crinkly detail looks pretty much the same.

This is characteristic of photographs of the moon. Craters come in all sizes, so it is hard to get a sense of scale. Even on Earth, where gentle rains and greenery erase the scars of planetary trauma, scientific photographs of rock formations often include a measuring stick for scale; otherwise it might be hard to judge the size. Mandelbrot said that fractals are all around us. They are the rule, not the exception.

Gott's Copernican method works when our knowledge of a duration has this fractal-like uncertainty. That is, we don't have a sense of the overall time scale; we don't know whether a measured past duration is a large or small part of the whole.

This does not apply to a seventeen-year cicada, whose time scale is conveniently disclosed up front. It does not apply very well to the lifespan of a dog or a human. Scale invariance better describes the lifespan of an amoeba. Amoebas can divide indefinitely. They are potentially immortal, though they can be and often are killed by unfavorable environmental conditions.

The scope of fundamental disagreement between Caves and Gott is less than might be imagined from the snark. Caves is saying that many processes are not scale-invariant (correct) while Gott is saying that many are (also correct). In 2008 Caves conceded as much, writing, "When you can't identify any time scales,

Gott's rule is your best bet for making predictions of a future duration based on a present age."

Brad Pitt's Wallet

"Estimate how much cash is in Brad Pitt's wallet." This, a recent internet challenge, was triggered by a news story giving the actual amount in the actor's wallet. You may want to formulate your guess before reading on.

It wouldn't be too hard to guess Brad Pitt's approximate age, weight, height, or credit score. These measurements have a characteristic scale. We know most adult men are about six feet tall. No reasonable guess about Pitt's height is going to be way out of the ballpark. Uncertainty about the cash in Pitt's wallet is of a more profound sort. He might carry around a big wad of cash to finance a glamorous, movie-star lifestyle. Or, maybe celebrities don't handle their own money. Cash is for little people.

This raises the question of how we assign probabilities to a quantity of no known scale. This issue was at the forefront in 1994, when *Nature* published letters critical of Gott's doomsday article.

When you have no reason to favor any of a set of possible outcomes, all should be assigned the same probability. This rule of thumb is known as the *principle of indifference*. We routinely apply it to coin tosses and lottery drawings. The great Laplace described the principle of indifference. He did not bother to justify it or even give it a name. Laplace apparently believed it to be self-evident.

Almost any gambler, then or now, would agree. That's because gambling equipment is precision-engineered to embody the principle of indifference. All six sides of a die ought to land with equal probability. Otherwise honest gamblers demand new dice.

Away from the gambling table, indifference is a trickier concept. "Either the Loch Ness Monster exists, or it doesn't. Nobody is sure which, so the odds must be 50:50."

It's easy to spot the error. There is ample reason to believe that the monster is a myth (absence of skeletons or fossils, numerous exposed hoaxes, the improbability that a large creature could perpetually evade determined efforts at detection, etc.). We must not ignore the data and invoke indifference for a new roll of the dice.

This is not just a theoretical problem. It confronts real-world deciders who have never heard of the principle of indifference. In his famous wager Pascal said that no one can be sure whether God exists or doesn't exist. Ergo, both possibilities deserve to be taken seriously. Climate change deniers sometimes take the tack that the evidence is not conclusive, so public policy ought to assume that climate change is equally likely to be real or not. By the mid-twentieth century, John Maynard Keynes could write (sarcastically) of the principle of indifference: "No other formula in the alchemy of logic has exerted more astonishing powers. For it has established the existence of God from total ignorance."

Statistician Steven Goodman quoted those very words in his rebuttal to Gott's 1993 paper. But, as Gott countered, Keynes himself mentioned cases where the principle of indifference is properly used. One is when a point lies at an unknown position on a line. The Copernican method says the present moment lies at an unknown position on a timeline.

It is one thing to use the principle of indifference when outcomes are easily identified, like "heads" or "tails." What do you do when an outcome can take on any of a wide range of numerical values? The most common answer is to use a *uniform logarithmic prior,* or a *Jeffreys prior.*

Sir Harold Jeffreys (1891–1989) was a British polymath who played a major role in the revival of Bayesian probability. Jeffreys proposed that, for unknown numerical quantities, the probability of every power-of-ten range should be the same. For instance, the chance of having between $1 and $10 in a wallet would be the same as the chance of having between $10 and $100, or between $100 and $1,000.

A good way to visualize that is to imagine throwing a random dart onto a logarithmically scaled number line. That's one where each power-of-ten range is the same size. Consequently, the chance of the dart landing between $1 and $10 is the same as the chance of it landing between $10 and $100, or in any other tenfold range.

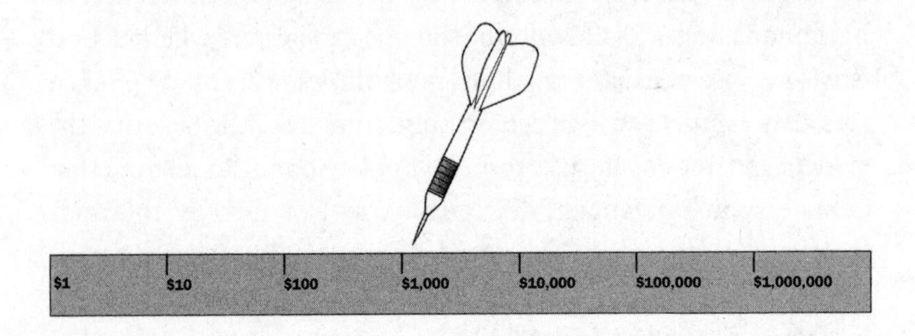

Apply this to Brad Pitt's wallet. We might naively think that, since we have no idea how much money is in Pitt's wallet, every possibility is equally likely. The trouble is that there's an infinity of whole numbers, and there are far more large numbers than small ones. This would lead to absurd conclusions, such as that Pitt probably carries more than $1 trillion (because most whole numbers are larger than a trillion).

Clearly, we can and must set reasonable bounds on how much walking-around cash Pitt has. Let's assume that he carries some money (at least $1) and agree to round the amount to the nearest dollar. The upper bound is set by how much currency is in circulation, how much of it a wealthy actor could possess, how much could fit in a wallet, and how much the actor would want to have in his wallet. Let's say the maximum realistic amount is $100,000.

We still don't want to say that every whole amount from $1 to $100,000 is equally likely. That would mean there's a 50 percent chance Pitt carries at least $50,000, which seems way too high.

Using the Jeffreys prior is like throwing the dart repeatedly at the logarithmically scaled number line and tallying how many times the dart lands within given ranges. In all, our bounds span five tenfold ranges. There's about a 20 percent chance of the dart landing between $1 and $10; another 20 percent chance of $10 to $100; and likewise 20 percent chances of $100 to $1,000, $1,000 to $10,000, and $10,000 to $100,000. The dart's median position is right in the middle of our bounds, a little more than $300. This probability distribution might be a reasonable model for our uncertainty about Pitt's wallet. "I think there's a 50:50 chance he has more than $300, and a 20 percent chance of his having a five-figure sum."

The Jeffreys prior expresses order-of-magnitude uncertainty when we don't know, even to a power of ten, what a value is. It is the only such probability function that is scale-invariant (for positive, unbounded quantities).

In his 1994 response to criticism, Gott showed that his Copernican method is equivalent to a Bayesian prediction using a Jeffreys prior for durations. The Copernican method is applicable to the extent that the Jeffreys prior is—that is, to situations in which we are truly, profoundly clueless about a duration.

Guinness Book of World Records

The Jeffreys prior motivated the wager that Caves offered Gott: "For each of the six dogs above ten years old on the list, I am offering to bet Gott $1,000 US, at odds of 2:1 in his favor, that the dog will not survive to twice its age on 3 December 1999."

Adult pet owners, and adult astrophysicists, understand that dogs almost never live to twenty. The big one-oh is not a random point in a dog's life.

"I just don't do bets," Gott told a *New York Times* reporter.

Caves replied, "It is inescapable that he doesn't believe his own rule in the case of the dogs."

In 2008 Caves checked up on the dogs. All six were dead. He would have won each of the six wagers. An art collector, Caves regrets missing out on an easy $6,000, "enough to buy a very nice piece of Australian aboriginal art."

Gott says the proper test of the Copernican method is predicting the survival of a random dog—not a cherry-picked elder canine already pawing at death's door. Most of the world's dogs are not ten years old. A random dog does have a good shot at surviving to twice its present age.

Caves structured his wager around the median delta *t* prediction, a 50 percent chance of a ten-year-old dog surviving at least ten more years. When you adopt confidence levels much higher than 50 percent, Gott's prediction is likely to be right, even when the scale-invariance condition is not met. For instance, Gott's math predicts that a ten-year-old dog is 90 percent likely to die between the ages of 10.53 and 200 years. Guess what? Most do.

The lifespan of a dog has an approximate maximum but no minimum. There is always the risk of a dog being hit by the proverbial bus. Or an actual bus. There is no counterbalancing luck that occasionally preserves a dog decades or centuries beyond its natural lifespan. Copernican predictions for an aged dog sound reasonable at the lower limit but overcautious at the high end. Indeed, Caves objected that "the intervals that Gott finds for survival times are so wide that he is likely to be right."

An elderly colleague of Gott's joked, "I'm so old I don't buy green bananas." He told Gott, also jokingly, that the Copernican method could not predict *his* longevity.

It did.

"I went to the *Guinness Book of World Records*," Gott said, "and I found the oldest person in the world. Her name was Jeanne Calment." She lived in Arles, France. Her dad had sold canvases to Van Gogh.

In telling the tale Gott recalled the precise numbers and dates. "She was 118 years old at the time my paper was written, OK? She had been alive for 43,194 days. I would predict [at 95 percent confidence] that she would last at least 1,107 more days or less than 1,684,566 days. Realize she's a special person! She's very unusual in age; it shouldn't work for her at all."

Gott's prediction was, "She'll live at least until June 7, 1996—that's three years—but she'll die before June 29, 6605.

"She died on August 4, 1997." Gott beamed. "I won."

How much money is in Brad Pitt's wallet? A 2012 *People* magazine item reported that the actor donated all the cash in his wallet—$1,100—to London's Southampton General Hospital. Pitt had been in England to shoot scenes for a zombie apocalypse film in Tunbridge Wells.

Baby Names and Bomb Fragments

Adolf Hitler proclaimed a thousand-year Reich at a September 1934 rally. Hitler had been in power a mere twenty months. A Copernican would have predicted the Nazi state to survive somewhere between another two weeks and another sixty-five years (at 95 percent confidence). The Third Reich lasted another eleven years.

Count that as a win for Copernicanism. One might think the date of doomsday is a matter of such utter uncertainty that no one could fault a method taking that uncertainty as an axiom. J. Richard Gott has found, however, that many are prepared to assert that we are, in fact, extraordinarily early in the timeline of existence. It's springtime for humans.

For Gott the Copernican method is a testable hypothesis. He has forecast not only the runs of plays but the ruins of celebrity marriages. The date of Di and Charles's royal divorce fell within the 90 percent confidence level. So did Gott's 1996 prediction for when the long-suffering Chicago White Sox would again win the World Series. (They did in 2005, not having done so since 1917.)

Gott showed me a memento, a 1981 desk calendar illustrated with photos of world monuments and wonders—the Eiffel Tower, Machu Picchu, Mount Fuji, and so on. All the famous sites existed in 1981. One no longer does.

The Copernican method says that a monument's future is likely to be in proportion to its past. It follows that the most recently constructed monument, at the arbitrary moment of the calendar's publication, is one you should bet on to fall first. First in, first out. Gott turned the page to a picture of the youngest of all of 1981's wonders of the world. It was New York's World Trade Center. It opened in 1973 and was felled by hijacked planes in 2001.

These are compelling stories. Gott is a great storyteller. Is there anything there beyond the anecdotes?

I will attempt to survey the methodical data bearing on the Copernican method. Because that method is a way to predict "everything," the relevant evidence spans many subjects and disciplines. It includes records of the survival of business enterprises, a topic on which business schools and economics departments have assembled large data sets (not to test the Copernican method, of course). This chapter will also offer brief detours into archaeology, philology, and Harry Potter.

But first, let's start with plays. In his 2009 book, *Apocalypse When?*, physicist Willard Wells went beyond Gott's Broadway experiment. Wells discovered J. P. Wearing's *The London Stage: A Calendar of Plays and Players*. This multivolume reference attempted to document every theatrical production in London from 1890 to 1959.

Set your time machine to random, with spatial coordinates for London's West End. We touch down and buy a copy of the *Times*. The masthead says the date is January 9, 1926. *Charley's Aunt; Peter Pan; No, No, Nanette;* Ibsen's *A Doll's House;* and Shaw's *Saint Joan* are playing. The chart below represents these and the

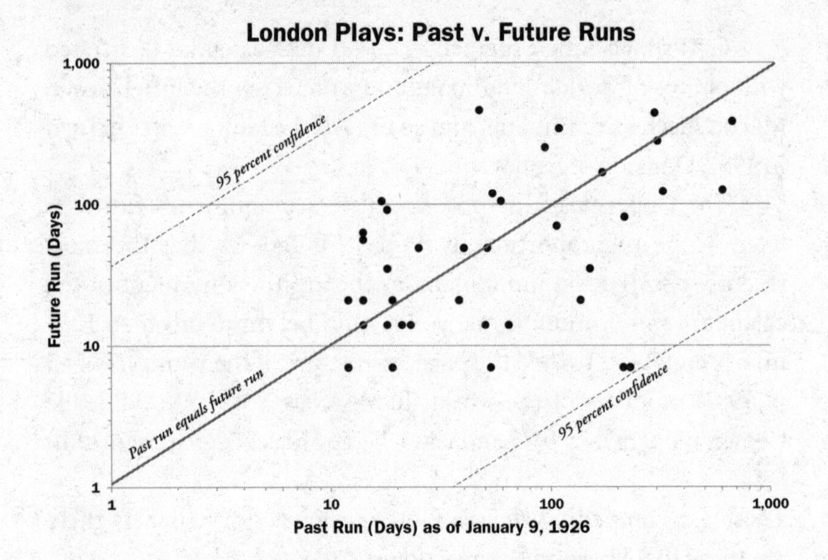

other plays then running with a dot. Each dot's horizontal position indicates a play's past run (how many calendar days had elapsed since opening night, in the then-current production). The vertical position indicates the play's future run. Because the data spans a couple orders of magnitude, I have used a logarithmic scale on both axes.

The bold diagonal line represents the median Copernican prediction, that a play's future run will equal its past run. It bisects the cloud of dots. Two dashed diagonals on either side of it show the 95 percent confidence limits. All the visible dots fall within these limits. (One dot is omitted, for *The Bohemian Girl*, a show that closed on our random night of January 9, 1926.)

When there is a strong correlation between two variables, a scatter chart's dot cloud compresses to a line. There is no such correlation here—zilch! It's random data, or nearly so. Gott makes a much more modest claim, that a random performance is unlikely to be extraordinarily early or late in a show's run. In terms of the chart, that means that few (in this case, no) dots fall in the triangular zones at the upper left and lower right. In order for a dot to land in the upper left corner, our random time machine would have to touch down very early in a play's run. For the lower right corner, the random date would have to be very late.

We can chart the same set of data in another way. Suppose we rank the plays by their total run (past plus future). By this criterion, the number one play running January 9, 1926, was *The Farmer's Wife*. This production ran for 1,054 days total. The numerical rankings are charted on the horizontal axis, and the length of the run on the vertical axis.

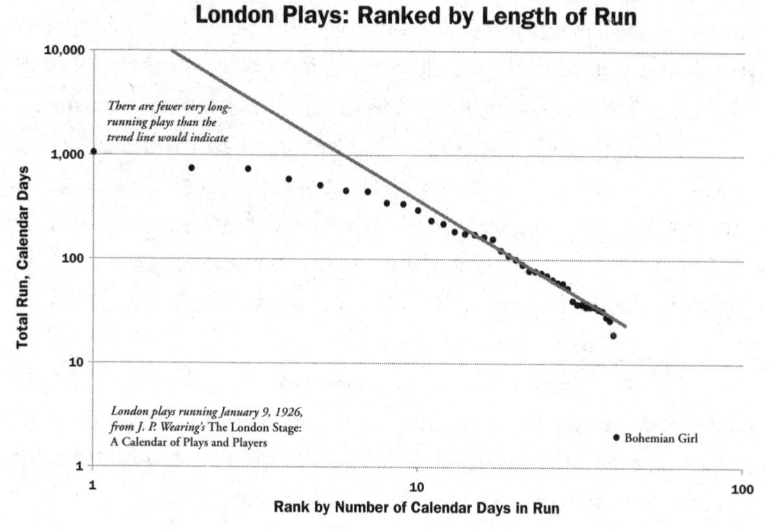

Now there's a proper line. Aside from *The Bohemian Girl,* the majority of dots follow a straight trend line on this log-log plot. There are, however, fewer very long-running plays (upper left) than the trend line would project. The longest-running shows tend to end sooner than would otherwise be expected (fall below the trend line).

Wearing documented 16,000 performances of Agatha Christie's *The Mousetrap*. At this writing, it is still in production, with more than 26,000 performances, and "looks unlikely to ever stop," the *Telegraph* claimed in 2015. But there are fewer such hits than the trend line would indicate. The point of inflection is generally somewhere around 250 days. Plays lasting no longer than that—about 85 percent of all productions—follow a line. Plays running longer than that are less common than predicted.

This isn't hard to understand. The Copernican model assumes that plays are timeless (scale-independent) and can potentially run forever. They aren't, and they can't. One time scale is provided by the population of Londoners and frequent tourists who regularly go to plays. After eight months or thereabouts, most who intend to see a given show have done so. Hard-core fans aside, most people are not going to see the same play twice. With the pool of likely ticket buyers dwindling, it becomes harder to stay in business.

Should the play surmount that hurdle, it confronts changing tastes. Opened in 1952, *The Mousetrap* has been a period piece for most of its run, to some degree insulated from fashion. But its manor-house murder is getting creaky for audiences raised on profanity-laden TV crime dramas. At some point plays grow out of sync with culture.

We may think the "timeless" works of Shakespeare are an exception. But they're a unique exception, and even there the language is an issue for today's audiences. Many productions of Shakespeare strive so hard to be contemporary that they are best described as original adaptations.

Gott discerns a cockeyed optimism in Broadway ads. The tagline for *The Phantom of the Opera* was "Eternally Yours." For *Cats* it was "Now and Forever." These mottoes stand in contrast to the typical run of a play. Everything humans begin must end, but we have a way of convincing ourselves that the statistics apply to other people, other enterprises. We all think we're special. Most of us are wrong.

Zipf's Law

George Kingsley Zipf (1902–1950) was a Harvard linguist whose nerdish obsession was the relative frequencies of words. Living before the age of computers, he used some of his family fortune to hire minions to count the occurrences of words in magazines, books, and newspapers. Zipf established that the most common word in the English language is "the." It accounts for about 7 percent of written text.

More important, he devised what's now known as Zipf's law. This says that the frequency of a given word is inversely proportional to its rank in the list of most common words. If the rank is *N*, then the frequency is proportional to $1/N$. That means a word twice as high up on the list of commonest words is about twice as common. The number one word, "the," occurs about twice as frequently as the number two word, "of," making up about 3.5 percent of English.

Zipf felt he was onto some deep and mystical truth. It is not just a quirk of English. It appears to apply to all natural languages and to some not-so-natural ones, like Esperanto. Zipf's law describes lists of the most populous cities, the most popular baby names, the most profitable corporations, and the most popular TV shows, as well as wars ranked by body count, and bomb fragments ranked by their size. Zipf's law governs wealth inequality, and it rules the internet, applying to rankings of most-visited websites and most-searched-for keywords.

A cottage industry developed for finding applications of Zipf's law. Until recently no one seems to have thought to apply Zipf's law to durations, but when it is, Zipf's law is very closely related to the Copernican method. Both are consequences of scale invariance.

The difference is that Zipf's law predicts durations from their position on a ranked list, while the Copernican method predicts future duration from past durations. But when you rank durations and chart them against their rank, as I did with the London plays, you get a Zipf's law trend line. Under the hood, Gott's and Zipf's rules share the same statistical engine.

Corporate Survival

It is the "Lindy effect" (or Lindy's law) that has become a business buzzword. Companies, markets, and managers with longer pasts are likely to have longer futures.

In a 2004 article José Mata and Pedro Portugal tracked the survival of Portuguese business firms for the decade beginning 1982. The Iberian nation required every business with even a single employee to report statistics. This resulted in an unusually complete data set making it possible to track how long businesses of all sizes lasted in 1980s Portugal.

The authors started with more than 100,000 firms and tallied how many were still in operation year by year. They found a simple curve, steep at first and then progressively slower. The median age of a Portuguese business founded in 1982 was 4.2 years.

That is not unique to sunny Portugal. Below is a similar chart of the almost 570,000 US companies founded in 1994, the same year as Amazon. The heights of the chart's bars trace a remarkably smooth curve. Most companies founded in 1994 are already out of business, and the median lifetime was about five years, similar to Portugal's.

In outline this isn't surprising. The number of surviving people in your high school or college class, or those born the year you were, can only decrease with each passing year. But the chart's line tells a somewhat different story. It gives the probability of a survivor company in a given year making it to the following year. This

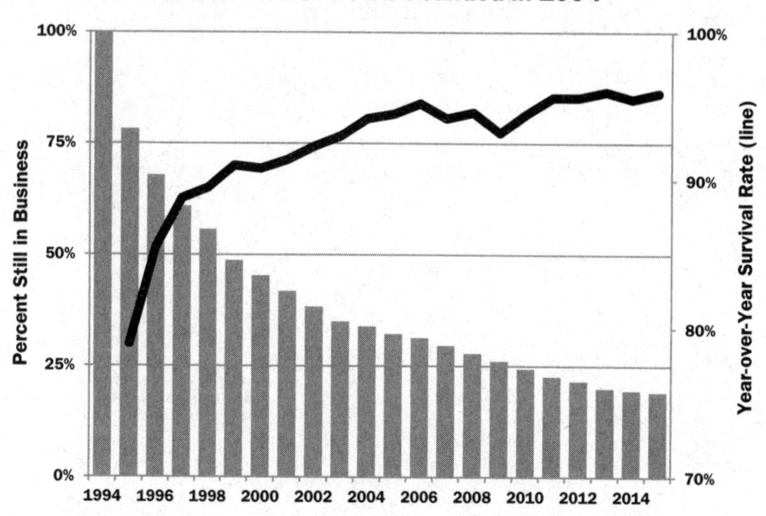

probability *increases* with time. The increase is steep in the early years and more gradual later on.

Human (and dog) demographics are not like this. A person's chance of making it to her next birthday decreases with each passing year. But companies are not like people or dogs; their chance of making it to their next birthday generally increases. It never reaches 100 percent, though it's over 95 percent for a twenty-year-old company in this set of data.

This is what the Copernican method (aka Lindy's law) predicts: a company's future duration increases apace with its past duration. Intuitive recognition of this phenomenon must motivate the practice of restaurant and business signs touting an "established" date. The subtext is that a business with a past is a business with a future (and is probably doing something right). Going by the companies' track records, it's likely that people will be drinking Coca-Cola (founded 1886) long after McDonald's (1955) has served its last hamburger, and the last Amazon (1994) drone has delivered its final package.

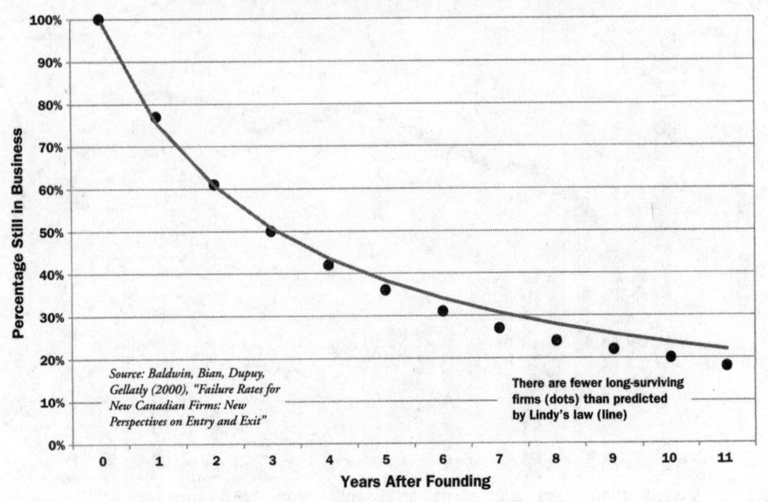

The chart above tracks the longevity of newly founded Canadian industrial companies. In this data set, the average start-up went belly-up in about three years. The fit to the Lindy prediction (curved line) is impressive, particularly for the first few years. As with plays, the data points sag downward from the line toward the right. There are fewer long-lived companies, though the effect is not so pronounced as it is for plays.

A business firm may have more opportunity or incentive to reinvent itself than a play does. But products and markets do become obsolete, and it is difficult for a company to survive revolutionary changes of consumer tastes or economic regime. Wells notes that, in the absence of obsolescence, the Lindy effect would predict the survival of a few companies from ancient times. While no such companies do survive, there's at least one major industrial corporation with roots firmly in the Middle Ages: Sweden's Stora Kopparberg is documented as having been in business by 1288. Originally a copper-mining firm, it moved into lumber and paper.

After a 1996 merger it took the name Stora Enso, and it is still in business.

Lindy in the Stock Market

John Burr Williams was a Harvard economist who sought to understand the 1929 stock market crash. Stocks had not been worth what investors of the Roaring Twenties thought they were. In *The Theory of Investment Value* (1938), Williams maintained that the value of any asset resides in its future income stream, discounted to present value. This is known as the discounted cash flow model.

According to Williams, a dividend-paying stock is worth the sum of all its future dividend payments, suitably adjusted. Suppose Coca-Cola pays a dividend of $1.59 a share and will pay that same yearly dividend for the next hundred years. The total would be $159. These dividends must be adjusted to reflect the time value of money. A $1.59 dividend now is worth more than a $1.59 dividend to be paid a hundred years from now. The difference is expressed in a discount rate that factors in inflation, risk, and opportunity costs. You would find that the value of a share of Coca-Cola ought to be a good deal less than $159.

What if a stock doesn't pay dividends? There are two answers. Either the investor intends to hold the stock a long time and expects it to pay dividends in the future; or (more likely) she expects to sell the stock at a profit. To put that into Williams's framework, you can consider any future capital gains to be a big and final "dividend."

Williams's model is a supremely rational method of valuing stocks. It can seem quaint, even irrelevant, in a world obsessed with buzz-worthy start-ups. Yet almost no one disputes Williams's basic idea. You buy a stock or property to make money; the more money you get, and the sooner you get it, the better.

This leads to a provocative conclusion. Some financial writers believe the Copernican method offers a way to achieve superior investment returns. As Lindy's law it is credited for the success of value investors such as Warren Buffett.

The premise is that an investor buys a stock or other financial asset at a random point of its existence. The asset's expected future lifespan is proportional to its past. This presumably applies not just to corporate longevity but to attributes of more direct interest to investors, such as revenue, earnings, and dividend payments.

Coca-Cola has paid dividends since 1920 and has increased those dividends each year for the past fifty-five years. Lindy's law predicts that Coke is likely to pay dividends, and increase them, decades into the future. It is more likely to do this than a company with less of a track record of dividend payments.

You may feel this is simply common sense. But it gets more interesting when you consider the long-standing puzzle of how to value stocks. Accept discounted cash flow, even as an inexact rule of thumb, and you discover something remarkable. The future lifespan of a company ought to matter in stock valuations, and it ought to matter a lot.

Ben Reynolds, who has examined Warren Buffett's investing in the context of the Lindy effect, gave this example. Suppose a stock pays a dividend of $1 a year now, and this dividend grows 6 percent a year until the company abruptly goes bust and its stock becomes worthless. You have a discount rate of 7 percent, meaning, say, that you can get 7 percent as a risk-free return elsewhere. With discounted cash flow, here's what the stock ought to be worth to you right now:

- $9.47 if the stock survives another ten years before going bust
- $18.03 if it survives another twenty years
- $39.10 if it survives fifty years

- $62.76 if it survives one hundred years
- $100 if it survives forever

In Reynolds's example the dividend grows almost fast enough to make up for the time value of money. So a stock that pays ten years of dividends before fizzling out is worth just a little shy of $10.

With longer corporate lifespans, the current value becomes progressively less than the sum of the future dividends. A hundred years of future dividends has a present value of $62.76, rather than $100. It would take an infinite future of dividends to be worth $100 to an investor right now.

No one can predict the next quarter, much less dividend payouts of coming decades. Fair enough. Lindy's law assumes the near-complete unpredictability of corporate lifespans. It is that unpredictability that leads to the prediction that old companies as a group have a longer future than new companies as a group. As we've seen, this prediction is supported by data. So, by discounted cash flow, the valuation of long-lived companies like Coca-Cola ought to be multiple times that of the latest start-up or IPO.

They're not! The highest-valued companies are often new ones that are burning cash. They have no earnings, pay no dividends, and have short track records. Investors think these new companies represent the future and have limitless upside. Established older companies are considered to be at risk of obsolescence. Lindy's law proposes the exact opposite.

It is no easy thing to distinguish talent from luck in the market. Warren Buffett is nonetheless among the most cited examples of presumptive stock-picking skill. From 1965 through 2013, Buffett's holding company, Berkshire Hathaway, averaged a compound return of 19.7 percent a year on its stock holdings, twice that of the S&P 500 in the same period (9.8 percent).

Here are Berkshire Hathaway's ten largest stock holdings, as reported in 2018, ranked by an unusual metric—how long the companies had been in business:

1. Bank of New York Mellon (234 years)
2. American Express (168 years)
3. Wells Fargo (166 years)
4. Kraft Heinz (149 years)
5. Coca-Cola (132 years)
6. Bank of America (113 years)
7. Moody's (109 years)
8. Phillips 66 (100 years)
9. US Bancorp (49 years)
10. Apple (42 years)

Buffett buys *old* companies. If his portfolio were a movie, it would be the kind where they bring back aging stars for one last geriatric heist. The median age of Buffett's top-ten companies is well over a century. By comparison, the median age of the ten biggest companies on the S&P 500 is only about forty-two years. (In 2015 Heinz, founded in 1869, merged with Kraft, founded in 1903. Buffett told shareholders that he expected the ketchup and pickle maker to "be prospering a century from now." Bank of New York Mellon is the creation of a 2007 merger, but both of its predecessors are venerable. The original Bank of New York was founded by Alexander Hamilton and Aaron Burr in 1784.)

Buffett's pithy sayings emphasize a long-term approach to investment. "Time is the friend of the wonderful company, the enemy of the mediocre." "If you aren't willing to own a stock for ten years, don't even think about owning it for ten minutes." "Our favorite holding period is forever." "At Berkshire, we make no attempt to pick the few winners that will emerge from an ocean of unproven enterprises. We're not smart enough to do that, and we know it. Instead, we try to apply Aesop's 2,600-year-old equation to opportunities in which we have reasonable confidence as to how many birds are in the bush and when they will emerge."

Aesop's "equation" is of course "A bird in the hand is worth two in the bush." For the most part, Buffett's birds-in-hand are long-established companies that—by Lindy's law and Buffett's research—are likely to have long futures.

A Copernican outlook challenges some of the venerable rules of investing. One is that stocks outperform bonds in the long run. This rule deserves to come with an asterisk, says Hendrik Bessembinder of Arizona State University's W. P. Carey School of Business. Historically, most stocks have done *worse* than US Treasury bills. That's allowing for reinvested dividends, capital gains, splits, *everything.*

How is this even possible? Bessembinder examined the returns of all stocks on the NYSE, AMEX, and NASDAQ exchanges from 1926 to 2015. He found that the stock market's performance is mainly due to a tiny minority of stocks that hit it big and remain successful. The average stock does far worse. The median life of a stock on the three exchanges was barely seven years. The most common return was close to a total loss.

If this is hard to believe, it's because indexes, like the S&P 500 or Dow Jones, do not say much about the typical stock. They are like Eddington's net, scooping up the bigger fish. All the stocks in the indexes are already winners. The indexes quickly drop companies that have stopped winning, replacing them with new winners.

Bessembinder's data justify one familiar rule of personal investing: buy index funds (mutual funds or ETFs that hold a portfolio replicating a market index, such as the S&P 500). Many investors disdain the indexes' returns as merely "average." The error of this view is generally demonstrated when a small investor tries picking stocks himself. A few randomly chosen stocks are likely to perform much worse than "average."

We've all heard of the chimpanzee that throws darts at the stock listings. The ape is usually cited to dismiss the value of expertise. Most professional stock pickers do no better than the "random" picks of an index fund. But that chimp is not in fact a good metaphor for index funds. Consider a different experiment. The chimp throws a dart at the front page of the *Wall Street Journal*. It buys whatever stock is mentioned in the editorial story that the dart hits. This will probably *not* be such an average stock. It's disproportionately likely to be a large-capitalization firm with many investors, employees, and customers. The *Journal*'s editorial policy is biased, naturally, toward companies with a large economic presence. Most of the firms that make the front page have already survived long past the average stock-issuing company's lifespan.

Another experiment would be for the ape to throw a dart at a listing of every share of stock—or better, every dollar invested in the stock market—and choose the company represented by that share or dollar. This is a better model of a capitalization-weighted index fund, such as one tracking the S&P 500. This form of random picking would also be skewed toward large-capitalization stocks that are, in some respects, not so average at all.

Bessembinder's findings imply that stocks are like lottery tickets. An investor needs to buy a large basket of them in order to have a fair shot at having a few winners—and thus matching the "average" returns we associate with indexes. Venture capitalists have a similar credo, the 1/*n* rule. As proposed by Benoit Mandelbrot, the 1/*n* rule says that it is better to divide one's capital into many shares and invest in many start-ups rather than just one. Most start-ups fail; most of a venture capitalist's return comes from a handful of rare successes. They're called "unicorns" for a reason.

Employees of start-ups often end up owning a lot of their own company's stock. Financial advisors warn against this, and Bessembinder's data shows why. Assuming the company is typical, its stock is likely to return less than T-bills. No sensible young investor would dream of having his life savings in T-bills, but many have them in an employer's stock.

As Buffett's history demonstrates, superior returns are possible just by buying companies with long track records of earnings. These companies have longer futures on average (says Lindy's law) and are, in rational economic terms, worth more (says discounted cash flow). The not-so-rational market often undervalues these companies.

Yet most investors set out to do the near-impossible: to identify the next big success in advance. They ignore the fact that most of the companies that have a real shot at being the next Amazon will be out of business in a few years. A new app could be released tomorrow that would make today's hot app obsolete or at least no longer so awesome. There is not the same risk of someone inventing a "better" soft drink tomorrow. Thousands of soft drinks have been invented in the past century, and for whatever reason none has been able to dethrone Coke. This is worth heeding, even if you've never understood the appeal of bubbly brown sugar-water. Investor-philosopher Nassim Taleb said of Lindy's law: "If there's something in the culture—say, a practice or a religion that you don't understand—yet has been done for a long time—don't call it 'irrational.' And: Don't expect the practice to *discontinue.*"

The Purloined Harry Potters

The survival curves we see for companies and plays are deeply ingrained in our world. Wells reported a similar decline in the number of copies of Harry Potter books in the San Diego Public Library. Too-ardent readers neglected to return the books. When entities have no known expiration date, similar statistics apply.

What (if anything) does this tell us about doomsday? At the very least, the fact that scale invariance governs corporate longevity, word frequencies, baby names, bomb fragments, stolen Harry Potter books, and Google searches ought to give us pause. We may be glimpsing a universal truth, one that has many aspects and has gone by many names: the delta *t* argument, the Copernican method, Zipf's law, Lindy's law, the Jeffreys prior, the doomsday argument...

We do not need to settle for data on plays and corporations. A galactic actuary, looking to sell extinction policies to *Homo sapiens,* might care more about our family history. There are about a dozen extinct species of *Homo* and related genera. In their day, they were the smartest creatures that ever walked the planet Earth. None of them lasted especially long as species go. And as Gott points out, hominid history is entirely consistent with Copernican predictions.

With the warning that these numbers cannot be known to great accuracy, given the luck of fossil hunting, here goes. Our own species has lasted about 200,000 years. Twelve other hominid species made it to the 200,000-year mark. They range from *Ardipithecus ramidus* (which lasted more than 250,000 years total) to *Homo erectus* (more than 1.4 million years).

The Copernican method predicts the median future lifespan for a 200,000-year-old species to be another 200,000 years. Six of the twelve species fell short of that, and six exceeded it. That's a bull's-eye.

At 95 percent confidence, a 200,000-year-old species should be good for another 1/39 to 39 times that. That would be 5,100 to 7.8 million years. All the extinct species fall well within this range.

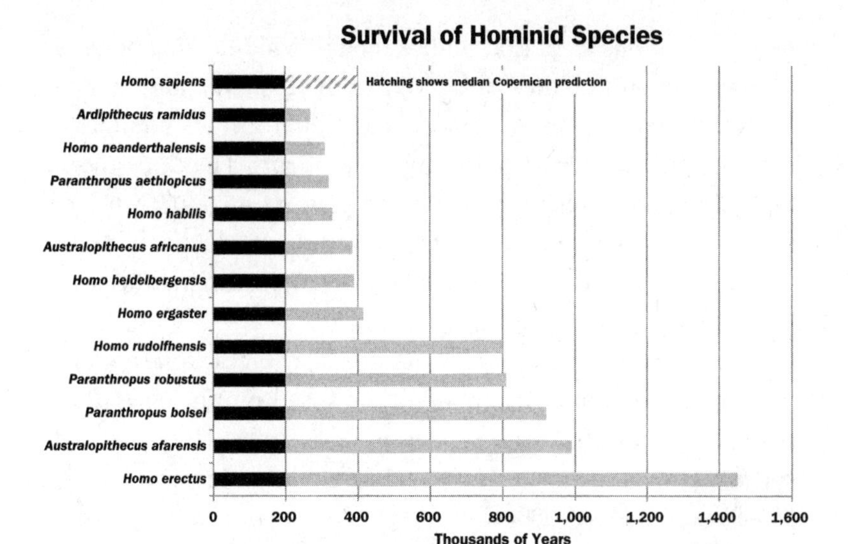

Gott calculates that about 68 percent of all hominid individuals would have found themselves to be of a species that had already existed longer than 200,000 years (had they been able to determine such a thing). Ours is still a young species. We may be destined to beat the odds, but so far we look average. To deny that is to assert a most extreme sort of *Homo sapiens* exceptionalism. ("I don't need to wear a helmet; I'm not like those *other* cyclists.")

It's true that *Homo sapiens* no longer has any competition. Some of the earlier hominid species competed against each other in a Darwinian elimination match, leaving *Homo sapiens* as the champion. It should be smooth sailing from here on out.... Except that no one really believes that. The death of the last Neanderthal did not put an end to strife. It just meant that we drew the battle lines a little differently, competing as nations, races, religions, and ideologies, possessing ever-escalating technologies of mass destruction. It is possible that nonhuman competition lies in the future. We may vie with unruly AI and, who knows, ETs.

To sum up, there is a modest though wide-ranging body of evidence supporting the Copernican method's predictions in matters other than the extinction of the human race. Sometimes a chart is worth a thousand thought experiments. The Copernican method applies to phenomena with no known time scale. Much of the criticism about it amounts to carping about whether Gott, or media coverage of Gott, spelled out this requirement in sufficiently large letters.

We can't test the Copernican doomsday prediction (not any time soon, we hope). Its credibility is based on analogies. The strongest parallel is to the extinction of our most closely related species. Though the data set is small, it fits the Copernican model as well as could be expected.

Yet those other hominids are not quite like us. Though *Paranthropus* had a survival instinct, it could not have grasped the concept of extinction. We are far cleverer, able to marshal our abilities to fight whatever threatens our collective being. Wells makes an interesting case that the play and corporation data may be more relevant than we think. Corporations and theatrical productions are microcosms. "All the world's a stage." They are made up of the same sort of fallible humans who will end our species or take it to the stars. These individuals have strong incentive to perpetuate their own and their groups' interests against a mix of internal and external threats. The group's time horizon is potentially limitless, even though it is composed of an ever-changing cast of mortals. These mortals have their entrances and their exits—by taking a gold watch, a golden parachute, or a Netflix original gig. They include stars, team players, bit players, rainmakers, and a few who will betray their colleagues the first chance they get. The company, repertory or limited liability, has challenges from without: high rents, high taxes, bad reviews, disruptive competitors, and the slow tick of a cultural clock that gradually renders anything less relevant than it once was. A madman's bomb or an orbit-crossing asteroid could end the most profitable run. A slightly bigger bomb or asteroid would end it all. Our species' future is dumb luck and office politics writ large: How to Survive in the Universe Without Really Trying.

Sleeping Beauty

You volunteer for an unusual experiment. On Sunday you take a pill that will cause you to sleep peacefully until it wears off three days later. After you fall asleep, a researcher flips a fair coin.

Heads, and you are awakened on Monday and interviewed. Then you are allowed to go back to sleep until the pill wears off.

Tails, and you are awakened on Monday and on Tuesday, and interviewed both times.

The sleeping pill causes temporary amnesia. In a given interview you won't be able to remember whether you've been woken up and interviewed before. You will, however, remember everything that happened before you took the pill, including the experimental setup. You will have all your usual powers of reasoning.

The experiment's interview poses this question: How likely is it that the coin toss came up heads?

This is "Sleeping Beauty," a brainteaser that became popular on the internet in 1999. It has become an integral part of the debate over the doomsday argument, specifically in its Bayesian, Carter-Leslie form. Sleeping Beauty and doomsday "are structurally the same," wrote Dennis Dieks, and "the analysis of one can be carried over to the other."

The Sleeping Beauty narrative has its roots in the "Awakening Game," a thought experiment devised by University College London philosopher Arnold Zuboff as early as 1983. In a 1990 paper Zuboff writes of "a game...being played in an amazingly big hotel" in which coin tosses determine the wakenings of drugged sleepers. Robert Stalnaker learned of Zuboff's work, supplying the catchier name "Sleeping Beauty." The riddle was transmitted through a social network of philosophy scholars in the Boston area, being distilled to a single sleeper and one or two wakenings. MIT grad student Adam Elga learned of Sleeping Beauty from Stalnaker and described it in a presentation at Brown University. Sarah Wright, then a Brown grad student, heard the talk and conveyed the puzzle to Brown philosopher James Dreier. It was Dreier who posted it on the rec.puzzles internet group on March 15, 1999. Elga was the first to publish Sleeping Beauty in a philosophy journal (*Analysis*), in 2000. The riddle now has a substantial literature. It encapsulates a central problem of self-sampling.

Opinions differ on whether the probability of heads should be one-half or one-third. Start with the halfer case, as it's called. It's a fair coin, and you know that. So the question is, do you learn anything in the course of the experiment that should cause you to adjust your initial belief that the chance of heads is 50 percent?

No. All you learn is that you're being awakened for an interview, as promised. This happens regardless of how the coin fell. You can't learn whether you're being awakened a second time, because you are unable to remember any earlier interview.

What part of "fair coin" don't you understand?

The thirder case starts with the observation that there are three wakening scenarios, all indistinguishable. Either it's Monday and the toss was heads (*but you don't know that*); or it's Monday and the toss was tails (*ditto*); or it's Tuesday and the toss was tails (*ditto*). Upon being wakened you have no reason to favor any of these three possibilities. Therefore, the principle of indifference applies. All three cases must be equally likely. But just one of the three has the toss coming up heads. The chance of heads is therefore one-third.

A thirder can take her beliefs to the bank. On the wall of your room is a sheet of paper that says:

WAGER

> The undersigned ("Bettor") wins $20 if the coin toss was tails and loses $30 if the toss was heads. (Sign on dotted line to accept this wager.) ______________

For a thirder, this is a good deal. On average, she expects tails (a $20 win) two-thirds of the time. She expects heads (a $30 loss) the other one-third of the time. This comes to an expected gain of $3.33. Were the experiment and bet repeated many times, a thirder would win an average of $3.33 per wager.

A halfer would reject the same wager. He expects a $20 win half the time and a $30 loss half the time, for a net loss of $5 per toss.

Who's right?

On this point there is consensus. If the experiment were repeated many times and the wager offered at every wakening, those taking the wager (thirders) would make money in the long run. Those refusing the wager (halfers) would be leaving profits on the table.

Halfers Defend Themselves

Thirders outnumber halfers. This is true of MIT grad students, brainteaser fans, and authors of peer-reviewed papers. As a minority group, halfers understand the majority better than the majority understands them. Halfers "get" the thirder case. Many thirders don't return the favor. They can't see how anyone could possibly be a halfer. They don't see why there is anything to discuss.

Let me then give a little extra attention to the halfer case, to show why it may not be so unreasonable. Start with the idea of repetition. Thirders say that one-third of all awakenings are heads. This is true (in the long run, assuming a much-repeated experiment). But halfers say that repetition of the experiment would prove that the coin lands heads 50 percent of the time. This is also true, of course.

Stuart Armstrong of Oxford feels that "probability is the wrong tool to be using" for the Sleeping Beauty and doomsday problems. Bayesian probability credits everyone with a coherent degree of belief for every proposition. Armstrong takes an approach more like a behavioral economist. We should focus on how people act rather than on what they say they believe.

I am awakened and asked for the chance of heads.

"One-half," I say.

"Maybe you'd be interested in a little side bet," says the interviewer, pulling out his wallet.

"Oh, no! You're going to exploit my amnesia and trick me into making a losing bet twice."

"I resent the implication that this important, serious experiment is some kind of scam!" (To himself: "What's wrong with a grad student making a little pocket change?")

The point is, I can believe the chance of heads is one-half and also understand how the experiment is rigged against me. Assume that I always bet on heads with even odds. There's a 50 percent chance of heads, in which case I'm offered and take a winning wager. There's a 50 percent chance of tails, in which case I'm offered and take a losing wager twice—on Monday *and* Tuesday.

That's how the interviewer cheats a perfectly reasonable halfer. (From the halfer's perspective, anyway. The interviewer may feel the bet "is what it is" and the halfer is wrong.)

There are remedies for the wary halfer. One is to bet *as if* the chance of heads is one-third. This need not be hypocrisy. It's just coming to terms with the peculiar features of the situation. The experiment's selection effect makes it appear to me that the fair coin falls heads one-third of the time.

Another remedy is to insist on a legal condition to the bet, stipulating that the wager can be offered only once per coin toss. Any additional wagers on the same toss are null and void. This would make the halfer odds correct and allow the halfer to place optimal bets.

Armstrong notes that the halfer-versus-thirder controversy also turns on ambiguities of personal identity. We are each a river into which we cannot step twice. Personal identity is a useful fiction, honed by evolution, to discourage us from doing unwise things. Before I take a leap into the void, I should reflect that the person who will lie dead at the bottom of the precipice is to be identified with the person at the top, who was crazy enough to jump.

But our gene pool never had to run a gauntlet of amnesia pills and confused identities. There, intuition may fail us. Writes Armstrong, "There are tricky problems distinguishing between 'I expect to see' and 'I expect someone exactly identical to me to see,' which can make degrees of belief ambiguous."

For instance: I wake up in the Sleeping Beauty experiment. I know Sunday's coin toss was fair, yielding a one-half chance of heads. This thought is complicated by the knowledge that there may be other awakenings of "me," indistinguishable from the present one. Only one-third of these awakenings will follow a toss of heads.

This becomes crucial when money is involved. It is usually

assumed that Monday-me will cooperate with Tuesday-me. We share not only the same name and DNA but the same wallet and credit cards. I will want to maximize the total amount won in wagers, by all my temporal avatars, as those winnings will go into the wallet that a future-me will take home on Wednesday. This leads to the thirder philosophy.

On the other hand, I might choose to live for today—not tomorrow or yesterday. *Carpe diem* isn't such a bad philosophy for an amnesiac. To make the point clearer, suppose that my wager winnings are to be paid immediately, in the form of a gift card that expires at the stroke of midnight. I can use the winnings to pay for movie rentals or gourmet takeout. All must be consumed that same day and can't be saved for tomorrow. This leads to the halfer philosophy. It too maximizes exactly what it claims to maximize. If I identify myself with *this awakening* only, then it follows that "I" can never be offered more than one wager. That guy who might be offered a bum bet tomorrow isn't me, and I don't care whether he's being swindled. It's no skin off my nose.

The Sailor's Child

Several variations on the Sleeping Beauty theme play with the role of personal identity. "Sailor's Child," devised by Radford Neal, replaces amnesia with another soap opera trope, the long-lost sibling.

It seems your father was an old-fashioned man of the sea, with a woman in every port. One night in a tavern he tossed a coin to decide whether to father one child or two. If two, they would be with different women in different ports.

You are his child, in Marseilles. You know the backstory but not the way the coin landed. What's the chance you've got a sibling in a distant port?

You are certainly a different individual from a possible half sibling you've never met, who would have grown up in a different culture. If you were placing a bet, you would probably want to "selfishly" maximize your own gain. You would likely not intend to share your winnings with the sibling, if there is one, or to donate the money to the Seamen's Fund for the Children of Stochastic Family Planning.

Neal, a thirder, believes that Sleeping Beauty's amnesia throws us a curveball. We are encouraged to think that the subject learns something in the course of the experiment and should update his beliefs. But no learning is possible with amnesia. In Sleeping Beauty, and more clearly in Sailor's Child, all one knows is the setup.

Duck or Rabbit?

There is no mystery about what happens in the Sleeping Beauty experiment. All agree the fair coin will land heads in one-half of repeated experimental trials. All agree that the experiment imposes a selection effect on the subject(s) such that heads precedes only one-third of awakenings. The halfer-thirder dispute is really over the best sound bite. "What is the likelihood of heads?" asks the TV reporter. (And keep it short!)

As thoughtful halfers see it, the challenge is to look beyond the experiment's blinders. Making allowances for the selection effect, what is the real, objective chance of heads? A fair coin has a 50 percent chance of heads. An amnesia pill doesn't change that.

Thirders see the question as an invitation to embrace the situation's novelty. That novelty lies in the selection effect, so they give an answer reflecting what I, and awakenings exactly like me, would observe.

There is no one right answer. We are being asked whether the drawing is "really" a duck or a rabbit.

**Welche Thiere gleichen ein-
ander am meisten?**

Kaninchen und Ente.

Shown is the first known publication of the duck-or-rabbit image, in an 1892 issue of a German humor magazine, *Fliegende Blätter.* Its artist went uncredited. The German caption asks, "Which animals are most like each other?" The drawing is usually called an "illusion," but that word supposes a unique reality that the drawing cleverly subverts. So it is with Sleeping Beauty.

The Presumptuous Philosopher

Niklas Boström was born in Helsingborg, Sweden, in 1973. Having spent most of his career in the English-speaking world, he anglicized Niklas and dropped the umlaut. Bostrom's great awakening came at age sixteen. He checked out a library book on German philosophy and went deep into the Nordic woods to read it. In a forest clearing, he encountered the ideas of Nietzsche and Schopenhauer.

This experience introduced Bostrom to the concept of an Übermensch. As described in Nietzsche's *Thus Spoke Zarathustra*, the Übermensch is a secular "Superman" transcending Christianity's focus on an afterlife and divine judgment. The Übermensch makes himself and the world the best it can be.

Since Nietzsche's time, the Übermensch has become a Rorschach blot. The Nazis spun Nietzsche to their purposes. So did American Jews of the time, who created superhero franchises that have long survived Hitler's thousand-year Reich. The Übermensch figures in the modernist canons of George Bernard Shaw (*Man and Superman*), James Joyce (*Ulysses*), and Alfred Hitchcock (*Rope*). Silicon Valley's most successful ride-sharing company, Uber Technologies, dropped the umlaut for business reasons.

Coming of age along with the internet, Bostrom became involved in the transhumanist movement. One of the first subcultures to be united by the global web, transhumanists meld Nietzsche, speculative technology, and science fiction into visions of the future. Among those visions is immortality. Transhumanists propose that people of the future may be able to upload their neural connections to computers and live forever as digital beings. It follows that it is important to survive to the point where immortality is possible. Many transhumanists plan to have their bodies frozen at death for reanimation by future technology. ("If you don't sign up your kids for cryonics then you are a lousy parent," said transhumanist Eliezer Yudkowsky.) Bostrom wears an ankle band with contact information for a cryonics facility in Arizona. As philosopher Daniel Hill said of Bostrom, "His interest in science was a natural outgrowing of his understandable desire to live forever, basically."

Another transhumanist tenet is the *singularity.* The term was first used by mathematician Stanislaw Ulam in 1958, recalling a conversation with John von Neumann (who died in 1957). In math, dividing by zero creates a singularity—a point where a function is undefined. Ulam and von Neumann used the jargon metaphorically, speaking of "the ever accelerating progress of technology and changes in the mode of human life, which gives the appearance of approaching some essential singularity in the history of the race beyond which human affairs, as we know them, could not continue."

Bostrom concluded that such heady possibilities were making traditional philosophy obsolete. An indifferent student until his discovery of Nietzsche, he mapped out a course of study that included physics, psychology, and computer science alongside readings in philosophy. As a first-year PhD candidate at the London School of Economics, Bostrom needed a dissertation topic. One day he saw a display of books at a conference. The title caught his eye: *The End of the World.*

It was John Leslie's book, and it became Bostrom's introduction to the doomsday argument. The idea "seemed interesting and important, probably wrong," he recalled. "I wanted to figure out why it was wrong because if it was *not* wrong it was really important."

Bostrom's advisors, Colin Howson and Craig Callender, "thought it was a crazy topic. Because I had two advisors I think each one was assuming that the other was paying a lot more attention to me and giving a lot of input." This gave Bostrom the freedom to pursue an unorthodox subject.

For someone who has spent much of his career studying the end of the world, Bostrom is funny. As a postgraduate he performed at open mikes in London comedy clubs. He also had exhibitions of his artwork. His wit and breadth of interests inform his dissertation, "Anthropic Bias: Observation Selection Effects in Science and Philosophy," which explores how to use self-sampling when selection effects exist. Harvard's Robert Nozick championed the dissertation's publication as a 2002 book. Ever since, it has been a key document of the contemporary philosophy of science.

What the Elephant Wrote

As stated earlier, the axiom of self-sampling is the self-sampling assumption, or SSA. Bostrom defines SSA this way: "One should reason as if one were a random sample from the set of all observers in one's reference class."

Note the "as if." I need not literally be a lottery drawing from all past, present, and future humans. I can reason as if I were, in order to draw conclusions.

We now need to decide what an observer is and what a reference class is. The first is pretty simple, but the second less so.

I, the elephant, wrote this.

—*What an elephant wrote in the sand, according to Pliny the Elder*

Observers are beings who use first-person pronouns. Pliny's first-century AD *Natural History* describes an elephant that had been trained to spell out Greek letters and words with its trunk. Pliny's elephant surely didn't understand the self-reference implied by dragging the tip of its trunk in the sand. So, are elephants observers?

Elephants are social animals that recognize their reflections in a mirror and appear to mourn their dead. But an observer is something more than that. It is a "Bayesian agent," as AI researchers say, one able to perceive the consequences of observations. There is no sharp threshold for that. But as a broad-brush generalization, humans are observers and elephants are not. Rocks and trees are definitely not. Intelligent machines could be observers, or so it's now widely assumed. Intelligent ETs would be observers.

The *reference class* is a group to which a particular observer belongs. More exactly, it's a group such that it makes sense to think of the observer as a random draw from that group. Don't let the convoluted definition throw you. In a lottery the reference class is clear enough. It is the set of people holding a ticket. I'm just a random person in that group.

But even in Sleeping Beauty, ideas about reference class can differ radically. Is my reference class just *this* awakening, or all of "my" awakenings, or all awakenings in the experiment, even of other people? The first leads to the halfer position, the second is open to interpretation, and the third implies the thirder answer.

For the doomsday argument the reference class is said to be the set of all past, present, and future humans. This may not be as clear-cut as it sounds. The witches tell Macbeth he can't be harmed by any man born of woman. Macbeth is slain by Macduff, who was born by Caesarean section. *D'oh!*

Today's transhumanists, and other visionaries, conceive of humanity as neither stable nor eternal. They contemplate a future of genetically or digitally enhanced humans—uploaded minds and purely artificial intelligences housed in android bodies or virtual worlds. The future may hold many other types of observers we can't begin to imagine now. To be human is to be in flux.

Suppose that, in the twenty-fifth century, uploading consciousness becomes a popular trend. Newly perfected technology is able to scan brains down to every neural connection, and to realize that person's consciousness in software as a virtual AI being. At first many wonder whether the uploaded entities are truly "alive" and whether the mental software has bugs. But the uploads *act* human. Immortality is a big selling point.

The rich and famous pay exorbitant prices to be uploaded. As the technology is perfected, the price drops, and everyone wants to be an upload. By the end of the twenty-sixth century, nearly everyone has gone virtual—or died. The last human "born of woman" has a birth rank somewhere around 200 billion. Not long afterward *Homo sapiens* is as extinct as the dodo.

Was the doomsday argument right? From today's perspective, biological extinction is a big deal. But it might not be the end of the world. Posterity could regard the adoption of uploading as an inevitable cultural trend, like the introduction of bronze casting, antiseptics, or mobile phones. Uploaded minds might see complete continuity between themselves and their biological precursors.

There was an end to the Han Dynasty, the reign of Richard the Lionheart, and the Great Depression. It remains to be seen whether the human race will end in a similarly clear-cut, recognizable way. Doomsday could be a judgment call even to those with a front-row seat.

If we count only biological humans in our reference class, then a prediction of early doom might not be so bad. It might just mean that human consciousness will soon take on new and improved forms.

But if we group all biological *and* uploaded humans (and their observer-moments) in our reference class, a forecast of doomsday would then have to be interpreted in more inescapably pessimistic terms, as the end of all flesh-and-blood and virtual humans, the end of anything resembling human consciousness.

This ambiguity is one problem with the doomsday argument and its reference class. Bostrom realized there are other, even more fundamental ones.

Musical Chairs

The *self-indication assumption,* or SIA, is a doctrine that was created by some very smart people for one reason—to make the doomsday argument go away. Like "impressionism," SIA is the coinage of an unsympathetic critic, in this case Bostrom. "When I started thinking about what's really wrong with the doomsday argument, [SIA] was the thing I came up with initially," he recalled. He wrote John Leslie about it, and "he was trying to talk me out of that."

In Bostrom's words the self-indication assumption says: "Given the fact that you exist, you should (other things [being] equal) favor hypotheses according to which many observers exist over hypotheses on which few observers exist."

Here Bostrom perhaps undersells SIA. Why should we put our thumb on the scale of evidence, arbitrarily favoring hypotheses with more observers?

SIA proponents say that such hypotheses have more "slots" or "openings" for observers. The fact that I exist is cause to believe in a large population of observers rather than a small one.

SIA is a game of musical chairs. There are a certain number

of chairs (slots for observers). When the music stops, each would-be player tries to sit down, occupying a random chair. But the players outnumber the chairs. It takes luck to stay in the game. And if a person who did get a seat is unsure how many chairs there are, the fact of having gotten a seat is reason to believe in a larger number of chairs.

Physicist and chess player Dennis Dieks articulated this basic idea in 1992. It's been discussed by Adam Elga, economist Robin Hanson, cosmologist Ken Olum, and many others.

We have already met a good illustration of SSA versus SIA. It is Sleeping Beauty. SSA leads to the halfer position, and SIA to the thirder position.

Remember, there are two hypotheses in Sleeping Beauty—heads and tails. Each possible awakening is considered an observer. Halfers (SSA) say that there is either one observer (on Monday) or two (on Monday and Tuesday). There can't be three. I regard this awakening as a random sample of the one or two awakenings that are actually realized. My evidence of being awakened is equally consistent with either. There is no cause to adjust probabilities. The chance of heads remains what it was before the experiment, one-half.

Thirders (SIA) say that tails gives me twice as many opportunities to be awakened into consciousness. That is reason to favor tails. Since only one in three potential awakenings follows heads, one-third is the chance of heads.

Apply this to doomsday. Say I believe that the ultimate past-present-future human race will have had either 200 billion or 200 trillion members. I take my early birth rank as random (invoking SSA) and conclude that it is a thousand times more likely with 200 billion humans than 200 trillion. This is cause to favor early doom, given almost any reasonable prior probabilities of it.

But SIA says 200 trillion humans is a thousand times more likely than 200 billion—because the larger number offers a thousand times more chances for *me* to exist. SIA produces an equal and opposite probability shift that *exactly cancels out that of the doomsday argument.*

"As an application of Bayesian reasoning, the [doomsday] Argument is impeccable," Dieks wrote. The doomsday math (SSA) resisted all past attempts at refutation because it's correct. It's just not complete. Accept SIA, and doomsday was all a bad dream. The crack of doom may come tomorrow or a billion years from now, but the chances are exactly the same as we thought they were before we ever heard of the accursed doomsday argument.

My Aunt from Saginaw

SIA is fantastic news... *if* it's valid. Why should we believe it? Does SIA work in real life?

One day my aunt from Saginaw comes into town for an unexpected visit. She wants to see a play, so she goes to the discount ticket booth and comes back with tickets for the two of us. There's no accounting for my aunt's tastes. For all intents and purposes, I'm seeing a random play! As we settle into our seats I leaf through the playbill and learn it's the nineteenth performance. Can I draw any conclusion about its future run?

Someone thinking along the lines of Dieks might say no. A long-running play sells, or will have sold, more tickets than an unsuccessful one. I therefore have more chances to buy, be given, find, or steal a ticket to a hit. That is reason to believe I am more likely to see a hit. (And every hit has a nineteenth performance, so that doesn't tell me anything.)

On further reflection, much depends on how my aunt came by the tickets. There could have been only one play available at the ticket booth. Maybe it was a flop about to close, and the producers discounted the tickets in desperation. Or maybe there were two plays available, and my aunt tossed a coin to decide. The details matter.

The probabilities also depend on the overall statistics of theatrical runs. A hit has more performances than a flop, but there are many flops for every hit. It is not a mathematical necessity that a random ticket is likely to be to a long-running hit. It wouldn't be in a city where everybody has a trust fund financing an autobiographical stage piece about their dysfunctional family. Such shows could be so numerous that their collective audience would exceed that of the few commercial hits.

We've already seen that there is empirical evidence for Gott's Copernican method. It predicts the runs of plays reasonably well, without making any sort of SIA-style adjustment.

This is not so decisive here. Gott and Wells chose random nights in New York or London, then tracked all the plays that were running on those random nights. This included hits and flops, in whatever proportion they actually existed. That's a reasonable way to parse the data. It's not the only conceivable one. Had they, in effect, put all the world's used ticket stubs in a big jar and drawn them at random, then the drawing would have been biased toward hits that sold a lot of tickets and ran a long time. This would have reduced or nullified the observed probability shift. Once again, the details matter.

Whether there is a tendency for a random observer to be a long-lived species depends on the statistics of intelligent life in the universe. This is something no one on this planet knows. Do we live in a galaxy of one-night stands, where nearly all technological societies are vanity projects fated to close on opening night?

Theory of Everything

> Those Nigerian email solicitations are so unlikely to be for real that I never bother to respond unless they are offering at LEAST $50 million.
>
> —*Steven E. Landsburg*

Big numbers are powerful things. But they ought not compel us to believe every story containing them. Otherwise, every con artist need only add a few zeros to the numbers in his spiel.

"I thought some more and thought, *Well, there are big problems with this self-indication assumption,*" Bostrom said. He devised a tale to demonstrate what he felt was wrong with it, the "Presumptuous Philosopher."

Suppose that physicists searching for the ultimate theory of everything have narrowed it down to two rival conjectures.

Theory #1 says that the universe has a trillion trillion observers.

Theory #2 says that the universe is *much* bigger than that and has a trillion trillion trillion observers.

One of the two theories has to be right. A new supercollider experiment will reveal which.

"The experiment is a waste of money!" bellows the Presumptuous Philosopher from his easy chair. It is a waste because he already knows the answer. Theory #2 is a trillion times more probable because it postulates a trillion times more observers than #1 does. That means Presumptuous himself is a trillion times more likely to exist. Theory #2 has to be right. It's a simple application of Bayes's theorem and SIA. (QED.)

Bostrom's point, of course, is that his philosopher is out of his mind. Scientific disputes cannot be decided so easily. While the Presumptuous Philosopher is a cartoon, there are pressing scientific questions in which SIA leads to equally hard-to-accept conclusions. Cosmologists theorize a multiverse that is vastly larger than the universe we can see. A multiverse would have far more galaxies, stars, planets—and observer life-forms—than the visible universe alone. SIA "automatically" favors the multiverse theory, as in Bostrom's tale.

The strongest case against SIA is infinity. Many hold the multiverse to be literally infinite, with an infinity of observers. By presumptuous philosophizing, our world *has* to be infinite, for the

Bayesian odds are infinity to one in favor! But no one believes that argument holds water, nor should they.

It was Bostrom's intention to show that the Presumptuous Philosopher's claim is so ridiculous that everyone would fall into line and agree that SIA must be rejected. In this Bostrom was wrong.

"Sometimes it is Bostrom who is presumptuous," sniped Oxford colleague Austin Gerig. "We must be very careful when analyzing theories that posit the existence of unknown people."

Ken D. Olum, a Tufts University cosmologist, believes that the Presumptuous Philosopher *would* be justified in favoring theory #2, all else being equal. Olum qualifies that by saying that glib counterexamples with fantastically large numbers of unseen observers can be deceptive. In such cases, we ought to consider Occam's razor as well as SIA.

William of Ockham (1285–1347) was, like Thomas Bayes, an English theologian-philosopher who's become a cult hero to contemporary rationalists. "Occam's razor" (using an alternate spelling of the town) is the credo that we should not believe weird or complicated explanations without compelling evidence. This maxim is known to posterity through a seventeenth-century phrasing, "Entities must not be multiplied beyond necessity." Observers would count as "entities," so Occam's razor tells us to be skeptical of claims of huge populations of unseen observers.

Olum believes that both SIA and Occam's razor are useful rules of thumb. In certain troublesome cases like the Presumptuous Philosopher, Occam's razor might lead one to discount the effect of SIA and vice versa.

Still, it's not always clear how to apply Occam's razor (or SIA). What if a relatively simple theory, well grounded in evidence, predicts an infinity of observers? That's not a hypothetical question for today's cosmologists.

Tarzan Meets Jane

The theory of probability is, in the most profound way, only common sense reduced to calculus." So wrote Pierre-Simon Laplace in 1814. The historical record paints a murkier picture. Probability theory can be a quagmire with no firm footing. It has entrapped many of the smartest people who ever lived.

Gottfried Leibniz was the universal genius who invented calculus, independently of Newton, and who inspired Voltaire's consummate know-it-all, Dr. Pangloss. Leibniz believed that throwing an 11 with dice was just as likely as throwing a 12. (An 11 is twice as likely. This is now considered high school math.)

In two tosses of a fair coin, what is the chance of at least one heads? Mathematician and physicist Jean le Rond d'Alembert thought the answer was two-thirds. (It's actually three-fourths.) D'Alembert's error appears in the *Encyclopédie* of the French Enlightenment, which d'Alembert coedited.

These notorious errors did not involve complicated math. They could have been exposed by tossing a few dice or coins. The errors were the result of misleading assumptions that weren't recognized as assumptions (and which therefore weren't tested)—the result of not having invented clear ways to talk and think about these matters.

With luck today's disputes over self-sampling will be trivial for high school students of the future. Until then a certain caution—and humility—is in order.

In 2007 Dennis Dieks, the godfather of the self-indication assumption (SIA), did something most unusual in this debate. He was persuaded. In a paper titled "Reasoning About the Future: Doom and Beauty" Dieks sided with Nick Bostrom on the Presumptuous Philosopher. He agreed that it is wrong to claim, as a general principle, that a theory predicting more observers is more likely to be correct than one predicting fewer.

Dieks remained adamant that the doomsday argument is a fallacy. He reiterated an objection he had made as early as 1992, that the doomsayers are counting the same evidence twice.

You're bicycling to work one day and see a gigantic urn set out in the park. It must be for a street fair or something. The urn is labeled "100 Balls." As you zip by, you see a worker putting a red ball into the urn. That's the only ball you see.

Hmmm, you think, pedaling on. *It's an urn full of red balls. Or else the balls are all different colors.*

If all the balls are red, then the chance of observing a red ball is 100 percent. If the balls are assorted colors, then the chance of seeing a red one is much less. Bayes's theorem tells you to favor the all-the-balls-are-red hypothesis. It makes the evidence less extraordinary.

Do you see the problem here? The guess that "all the balls are red" was motivated by your glimpse of a red ball. Had you seen a green ball, you wouldn't be wasting your time with the all-red theory. You'd be weighing the all-green theory against the multicolor theory. The red ball sighting is already baked into the all-red theory. Writes Dieks, "If a certain piece of evidence has already been incorporated into a hypothesis, it is, of course, not permissible to treat that evidence as independent information bearing on the reliability of the hypothesis."

Dieks believes that we have unwittingly smuggled knowledge of our place in human history into our doomsday thinking. Suppose I try to distinguish between two hypotheses:

1. Human extinction will occur by AD 2500.
2. Humans will survive past AD 2500.

This wording looks neutral and objective. It says nothing subjective about my location in time. But I wouldn't be bothering with a hypothesis like "human extinction will occur by AD 1700" because I already know that's wrong. I have chosen a cutoff point in the future, relative to my *now.* In that roundabout way knowledge of my point in time constrains the hypotheses I consider. Dieks says that ought to prevent me from treating the current time or my birth rank as brand-new evidence. To use the doomsday argument properly, I would have to erase my brain, unlearning all I know about my place in time and history.

Forgetting the Century

I'll give two reactions to that. One is that, if all we want is jurors uncontaminated by facts, they're not so hard to come by. Ask the first twelve people you pass in the street: How long ago did human life begin? What is the cumulative population of the human race?

Sure, they will be able to recite what year it is—that is, years since Jesus—but our concern is births since Adam. On that the person in the street is uninformed.

Another reaction is that it is not so easy to be impartial. Numerous psychological studies show that apparently well-meaning people exhibit gender or racial bias without realizing it. The doomsayers face a similar challenge in trying to be chronologically unprejudiced.

To not know one's place in time is, as Dieks says, "a bizarre situation, very different from our actual one." He offers this demonstration. You are hypnotized and instructed to forget who you are and when you live. You could be a medieval pope or a futuristic hobo; you could be "Lucy" of fossil fame or sitcom fame. You remember all facts not pertaining to your personal identity, however, and are able to reason from them. You are somehow aware that people of the twenty-first century had/do/will have certain anxieties about the future. You can express these anxieties in numerical probabilities of doomsday, from the perspective of a person in the twenty-first century.

An unstated premise of the doomsday argument is that the usual prior probabilities of doom are the same ones you would have arrived at, had you not known your position in history. This Dieks disputes. To see why, imagine a super-simplified case in which a fair coin toss in AD 2500 decides the fate of humanity. Heads, and the world ends right then with a bang. Tails, and we go on to a long future with stupendous populations on many planets. The tails population is a thousand times that of the heads population.

Someone who understands this, and who knows that he lives before the fateful coin toss, must of course put the chance of heads at 50 percent. But since hypnotized-you doesn't have any idea of when you live, it's possible that the crucial year 2500 is already in the past. If so, that would rule out heads. This is reason to lower the likelihood of heads, relative to the estimates of your unhypnotized self. Specifically: given that there are a thousand times more people in the case of tails, you are a thousand times more likely to find yourself in the tails world than the heads world. Hypnotized-you should be very confident the toss was or will be tails. This, says Dieks, is just a matter of being consistent.

The hypnotist snaps her fingers. You suddenly remember what century it is. You discover that the coin toss is still in the future. This is a thousand times more likely with heads. But if you want to use Bayes's theorem to adjust the odds, you must use your naive probabilities, innocent of your location in time. The heads odds, which were cut by a factor of one thousand when you were hypnotized, must now be multiplied by one thousand. This cancels the doomsday shift, exactly as SIA would, but without that problematic assumption. You end up with the same prior probability you had before you were hypnotized (in this case 50 percent).

Bayes in the Jungle

Tarzan lives in the supreme isolation of the jungle. His only companions, the apes, are disconnected from the global news grid. Tarzan does not have a clue what year it is or where he fits into human history. Honestly!

Tarzan conceives three theories of human existence: small, medium, and large. Under the small theory, Tarzan is the only human being who has ever lived or ever will live. Under the medium theory, there is a vast human world beyond the jungle, with a past, present, and future population of 200 billion. Under the large theory, the cumulative population is 200 trillion. Tarzan believes all three theories are equally likely.

One day Tarzan meets Jane, a woman from the outside world. Jane teaches Tarzan history. She tells him it's the twentieth century, and about 50 billion humans were born before Tarzan was.

This (not to mention Jane herself) rules out the small theory. Otherwise Jane's news is consistent with either medium or large. Tarzan must now redistribute the one-third probability he formerly assigned to the small theory to the two still-viable theories.

But it shouldn't be apportioned equally. Tarzan's self-locating evidence, of having a birth rank of about 50 billion, is a thousand times more likely under the medium theory than the large theory. This produces a probability shift greatly favoring the medium theory over the large.

Tarzan seems immune to Dieks's critique. He did not import any historical knowledge into his original thinking because he didn't have any historical knowledge. Tarzan really did learn something from Jane, and it allows him to rule out the small theory. This in turn forces a redistribution of likelihood.

Our real-life situation is much like that of the educated Tarzan. We have learned things about human history that our ancestors—or younger versions of ourselves—didn't know. This rightly affects what we should now believe.

The Tarzan story has a distinctly implausible element (aside from a feral British viscount who wrestles gorillas). Tarzan is a perfect Bayesian.

He starts with a naive set of prior probabilities that must be taken as a given. Upon learning his place in history he adjusts these probabilities according to Bayes's rule. Tarzan's consistency can presumably be demonstrated, say by Dieks's experiment in hypnosis. But real people are not perfect Bayesians, nor are they perfectly consistent in what they believe.

Regular folks don't go around with a number (much less a probability distribution) in their heads representing what they believe about the end of the world. Any opinions on the matter are generated on demand. Should a pollster ask, most of us would mull the question over for a few seconds, rehearsing what we've heard about unstable world leaders, nuclear buttons, biohackers, and melting ice packs. Then we'd render an opinion that would track the median of opinions we've recently encountered, adjusted up or down for a current level of serotonins. Hardly anybody would do a Bayesian calculation, even if they're among the tiny minority who know what a Bayesian calculation is.

In practice "prior probability" describes all-too-human beliefs that are vague and inconsistent. Dieks is right in saying we shouldn't double-count evidence—but it can be difficult to tell what evidence has been counted.

The Shooting Room

"I spent two and a half years in absolute misery." This is John Leslie speaking of his obsession with the doomsday argument. Even while rock climbing with a partner, he would get ideas. He would have to stop, halfway up the rock face, to write them down.

"On two occasions, having woken in the middle of the night, working till dawn, I was actually physically sick. Because I hadn't seen my way through the paradox after working three or four hours." Leslie apologized to me that he knew little of recent publications in the field. He had, for his health, sworn off thinking about the doomsday argument—"because I was so near to ending up in the lunatic asylum."

During his period of intense engagement with doomsday, Leslie was corresponding with philosopher David Lewis. Lewis also had been waking up in the middle of the night, thinking about the doomsday puzzle. "There was a group of his grad students at Princeton who, for a full year, were discussing this over beer. And they weren't coming up with any decent refutations."

In response to some of Lewis's ideas Leslie devised a haunting thought experiment known as the "shooting room." It is a cartoon vision of doomsday, a tale of exponential growth that ends badly.

Lewis declared the shooting room "a good, hard paradox." It was published in Leslie's 1996 book, *The End of the World*, and has been a part of the doomsday conversation ever since. It offers a closer parallel to the doomsday argument than Sleeping Beauty does—and thus stands as a definitive demonstration of why people find it hard to agree about doomsday.

Abandon All Hope

A new game is beginning. The first prisoner is summoned to the shooting room. Its entrance is marked with a poem:

Abandon all hope, you who enter this room!
Well not quite all hope—and here's why:
You've a 1/36 chance of meeting your doom,
Yet 0.9 of those entering will die!

This does not look good, thinks the prisoner. (Poem courtesy of Paul Bartha and Christopher Hitchcock, by the way.)

That reading is quickly confirmed by the Commandant, who orders the prisoner to stand back, against the wall, so that the firing squad can get a good clear shot. Excellent! Then the merciful Commandant rolls a pair of dice. Should it come up two sixes, the Commandant will order the squad to fire. Otherwise the prisoner will be immediately pardoned and released unharmed.

The dice land 2 and 5. The Commandant waves the prisoner to the exit door, and the prisoner wastes no time in using it.

The moment that first prisoner leaves, the shooting room's walls pull outward a few feet. A door opens and nine new prisoners come in to take their place in the enlarged shooting room. Again the Commandant rolls the dice. If double sixes come up, he gives the order to fire on all nine. Otherwise, the nine are pardoned, and the room expands again, to accommodate a group of ninety.

As long as the dice land on anything but double sixes, there are ten times more prisoners at each iteration after the first—1, 9, 90, 900, 9,000, 90,000... The walls may expand, if necessary, to encompass the whole world. But eventually the Commandant will roll double sixes and everyone will die. Game over.

Imagine that you find yourself in front of the firing squad. What would you say your chances are?

The chance of rolling two sixes with fair dice is 1/36. Your survival depends on the dice and nothing else. It makes no difference how many prisoners came before you, nor how many come after. The obvious answer is that your chance of being shot is 1/36 or 2.78 percent.

Here's a less obvious answer. You are a random observer of this senseless game. At any given stage beyond the first, the number of people standing in front of the firing squad is 90 percent of the cumulative total of all who have done so. When the order comes to fire, and it will come eventually, 90 percent of all who had ever entered the shooting room will die.

(A minor exception is the case in which the Commandant happens to get double sixes on his first roll. Then the one and only prisoner dies, and the death rate is 100 percent.)

It appears reasonable to say that, as a random participant, your chance of being shot is 90 percent. That's a lot more than 1/36, and the difference is of no little interest. Which probability is correct?

Most feel that 1/36 is right. The prisoners have every reason to believe they will probably walk out unharmed. But note that an insurance company selling life policies to the prisoners would go broke by accepting the 1/36 chance. It would have to base its rates on the premise that 90 percent of all who enter the shooting room die.

It gets even stranger when you consider prisoner George and

his mother, Tracy. George is summoned to the shooting room. He texts Tracy to tell her not to worry too much. He's figured it out, and his chance of survival is 35/36. Tracy then drops her phone on the subway tracks and can't receive any more texts. When she gets home, she turns on the cable news. The chyron headline says ANOTHER SHOOTING-ROOM MASSACRE. Tracy can only believe the worst, that George is 90 percent likely to have been in the final, fatal round.

Our future is a series of existential crises whose outcomes we can't predict. The odds of surviving from day to day, century to century, are good. But one day our luck will run out. Meanwhile the human population tends to increase exponentially with time. Leslie's shooting room is a concise allegory of our hopes and fears.

Before going any further, I should mention the elephant in the (shooting) room. The numbers in this thought experiment aren't "realistic." On average the Commandant can expect to roll the dice thirty-six times to get double sixes once. A thirty-sixth round would demand a fresh group of 9×10^{34} prisoners. That's well over 10 trillion trillion times the current world population. The Commandant would need more than the world population just to advance to the eleventh roll.

Are we being overly literal in pointing this out? I would say so. The shooting room invokes exponential growth for the same reason that Thomas Malthus and Gordon Moore did. It's our lived reality, for the time being. A sizable fraction of all the people who have ever lived are living right now—in the last round?

So let's suspend disbelief on the numbers. Crowd-management issues aside, there is nothing magical or mysterious about the shooting room. Everything that happens is determined strictly by throws of dice. The odds of rolling dice have been understood for centuries. How can there be any dispute about the shooting room's probabilities?

* * *

Arnold Zuboff, of Sleeping Beauty fame, compares being improbable to being dangerous. Both are matters of perspective. We say a tiger is dangerous. What we mean is that a tiger *can* be dangerous. A tiger is not dangerous to me if it's in a zoo or on the other side of the globe.

The shooting room is certainly a dangerous place to be. But that doesn't mean we can measure that danger with a single number, representing the probability of death. Probabilities are contingent, based on states of knowledge and how that knowledge is used. Different people know (pay attention to) different things.

Tracy is treating the shooting room as a black box. She knows that 90 percent of all who walk in never walk out. She translates that fact into a statement of probability.

George is focusing on the dice. He knows his fate will be decided by the Commandant's next roll. Given how dice work, George concludes that he has a 35/36 chance of survival.

Ultimately George's answer is the more meaningful one. That's because George uses more information. Though George is aware of the 90 percent statistic, he has concluded that the dice throw provides a more direct and accurate way of calculating his chances.

Tracy is either unaware of the dice game's rules or is choosing to ignore them. If she truly doesn't understand the inner workings of the shooting room, then her 90 percent figure is justifiable—for someone in her incomplete state of knowledge. But it should come as no surprise that George has a better answer.

In fairness to Tracy, we all are prone to ignore information and settle for a quick, easy answer. This isn't always a problem. In more typical situations Tracy's random sampling would lead to the same answer as George's more detail-based reasoning, or nearly so. But Leslie has crafted a story in which the two approaches diverge sharply.

Möbius Strip

If you want to understand how a surface can have only one side and one boundary, make a Möbius strip. That little paper model of math classes demonstrates that what sounds like a paradox isn't. It's a real object you can hold in your hand.

The shooting room shows how the "same" event can have two probabilities—and why this isn't a paradox. Leslie distilled a central cognitive difficulty underlying the doomsday controversy into a story involving only rolls of dice, without bringing in the heavy emotional baggage of human destiny and the future.

There are several ways of unpacking the shooting room as a model of doomsday. Here's one: George corresponds to someone who understands all the underlying causes of human extinction and is able to supply accurate probabilities for them. Tracy is like someone applying the doomsday argument—specifically Gott's birth clock version. She ignores or does not have access to underlying causes. Her estimate is based entirely on random sampling from an exponentially growing population.

George's reasoning is much to be preferred to Tracy's. That's because shooting-room George really does know how the dice game works. Were we in George's position, able to analyze the probability of human extinction from first principles, there would be no need for the doomsday argument. We would already know all that can be known about the date of doomsday.

Yet it is doubtful that anyone understands doomsday risks so well. Not only are the risks difficult to assess, but humans are ever-resourceful. We may yet find ways to avoid global war, climate catastrophe, robot uprisings, or whatever lies ahead. This too must be factored in.

Accept this view, and it may not be feasible to set meaningful prior probabilities for doomsday. We would be like a Tracy who does not know the inner workings of the shooting room but is able to reason from statistics. The Copernican doomsday argument could offer the best estimate available in our limited state of knowledge.

Where does the Carter-Leslie doomsday argument fit in? It falls somewhere between George and Tracy. The Carter-Leslie argument assumes that we have beliefs about doomsday that are strong enough to be meaningful but not so strong as to render irrelevant the statistical clue we might get from our "early" birth ranks. It also requires that birth rank supplies information not already incorporated in our doomsday beliefs.

Leslie supplied a very different spin on the shooting room: it is about determinism. Upon seeing the TV news, Tracy knows that George's fate is sealed. At that moment there could be, in principle, a complete list of the game's survivors and victims. Tracy does not have this list but she has cause to regard George as a random draw from it.

To George, inside the shooting room, his fate is still up in the air. The crucial dice roll is unpredictable (to George, anyway) and has yet to happen. At that moment, no mortal could draw up a complete list of all the shooting room's past and future occupants. George has less reason to think of himself as a random draw from a list and every reason to focus on the dice.

The doomsday argument asks us to imagine a drawing from a list of all past, present, and future humans. In Leslie's view this list could exist in principle right now—*provided* the future is predetermined. The doomsday argument is compelling if the world is deterministic, with future events completely determined by the past. Otherwise, says Leslie, in an indeterministic world, the case for doomsday is weaker.

"The issue of determinism is a red herring," William Eckhardt countered. "Statistical inferences... do not hinge on the truth of determinism. That is why the determinism question is not a burning issue among say, insurance companies."

Neither insurance underwriters nor physicists are able to say whether the future is predestined. This leaves determinism as one of the longest-running of all philosophical debates. Quantum uncertainty and chaos theory limit our ability to predict certain events. No one knows whether this restriction is fundamental or only apparent, rooted in our imperfect knowledge of physics. Writes Eckhardt, "As long as the validity of the Doomsday argument is made to hinge on whether the future is open or fixed, or whether the future is fully implicit in the present, we can rest assured we are not going to settle the question of the argument's validity."

The Metaphysics of Gumball Machines

Good systems tend to violate normal human tendencies." That is the philosophy of a philosophical commodity trader, William Eckhardt.

Eckhardt left a nearly complete PhD in mathematical logic at the University of Chicago to join his high school friend Richard Dennis on the trading floor. Dennis was already a successful commodity trader who had turned about $5,000 into $100 million. Within a few years Eckhardt accumulated his own fortune. He traded futures contracts, meaning that he placed bets on the future prices of things like silver, cocoa, unleaded gas, and the Japanese yen. The actual commodity is the least important thing in commodity trading; all that matters are the ups and downs of its price. Successful trading is an exercise in behavioral economics, in predicting how fellow traders will fail to guess probabilities accurately.

Eckhardt and Dennis often debated the origin of their wealth. Was it brains, luck, or something else? Dennis believed it was an algorithm. Their system boiled down to a few rules. It had taken creativity and hard work to devise their moneymaking recipe, but now that it existed, it could be implemented by anyone off the street.

Eckhardt felt that trading required something more. It was not so much intelligence as discipline. Their system profited from the not-quite-rational market. To use it one needed the nerve to override powerful, hardwired instincts about money and risk. Someone without the right emotional makeup would not be able to trade successfully, even with the system.

Dennis and Eckhardt decided to settle the matter with an experiment. They took out ads in the *New York Times* and *Wall Street Journal* saying they were looking for people willing to learn trading. Over a thousand applied. Thirteen were accepted. They became known as the Turtles. (On a trip to Singapore, Dennis had seen a turtle farm. He said he intended to grow traders the same way.)

The group assembled in Chicago in December 1983 for a training course that took a mere two weeks. The Turtles began trading with real money in January. By February, most had been allocated $500,000 to $2 million of Dennis's money to manage.

The Turtle system (which has since been revealed in considerable detail) was intended to spot trends early, ride them up or down, and get out at a profit. It's reported that the Turtles made more than $175 million in five years. The experiment's success provided evidence for both Dennis's and Eckhardt's views. Though some Turtles made millions, a few did not. Some simply did not or could not follow the rules. The majority of Turtle system trades resulted in losses. The profit came from a few big wins. But these wins didn't just drop in the traders' laps. They required sticking with a position as its value fluctuated wildly. The trader had to overcome "normal human tendencies" to sell on strong downturns, to settle for a small profit rather than risking a loss. The unsuccessful Turtles bailed out early, missing the few big wins that made the system work.

In 1991 Eckhardt struck out on his own, founding the Eckhardt Trading Company, a commodities and alternative investments firm now managing more than $1 billion. With a shaved head and trim goatee, he has the Mephistophelean vibe of a hip card magician. He remains interested in the scholarly literature of probability theory and the philosophy of science. It was there that he encountered John Leslie's early publications on the doomsday argument. Eckhardt was sure that Leslie was dead wrong.

Despite not being a credentialed academic, Eckhardt has published influential articles on doomsday and the shooting room in *Mind* (1993) and the *Journal of Philosophy* (1997). To Eckhardt, there is no such thing as a probability paradox. There are only probability fallacies. He believes the Carter-Leslie doomsday argument falls into that category.

Urns Versus Token Dispensers

Eckhardt contests Leslie's urn analogy. Birth ranks are not random drawings. They are numbers assigned in serial order. George F. Sowers Jr. illustrates that with this tale. Your boss tells you to count the number of balls in a big urn. It's known that the urn contains either ten or one thousand balls. You start counting: One...two...three...You're up to seven when the micromanaging boss pops back into your cubicle. "Well? What's the answer?" he asks.

All you can say is that you haven't gotten far enough to tell a damn thing. That you're at seven *doesn't* mean the urn is more likely to have just ten balls.

Eckhardt suggests that we ditch the urn and replace it with a numbered-token dispenser. Picture it as a gumball machine with the glass top painted opaque. Once a minute the machine dispenses a gumball. As it does so, it imprints the ball with a serial number (in strict numerical order, which of course is not random at all). You see the machine dispense ball number 7. What does that tell you about the total number of balls in the machine?

This, says Eckhardt, is a better model of our fertile species and its existence through time. My birth was one of a long series of births that one day must end. But my birth rank is just a serial number, not a random draw from the whole set of once-and-future birth ranks. Therefore, learning my birth rank tells me nothing about how many people will exist after me.

In a 2009 article Paul Franceschi developed these ideas. Franceschi says we can imagine two kinds of numbered-token dispensers. One is compatible with the doomsday argument, and the other isn't.

1. The Carter-Leslie model. Somewhere inside the machine, a mechanical hand tosses a coin to decide whether to dispense ten or one thousand gumballs. The complete allotment of balls drops down from a hidden, soundproof reservoir before the first ball is dispensed. Then the machine dispenses its contents, one ball a minute, until it's empty.

Unlike a lottery urn, this machine has no random drawing. The randomness resides in my encounter with the machine at an arbitrary point in its operation. (Compare Gott's random encounter with the Berlin Wall.) Observing the machine dispense ball number 7 gives me strong reason to think there are only ten balls. This machine embodies the Carter-Leslie doomsday argument.

2. The Eckhardt model. The machine is filled in stages. First it's filled with ten balls. Then, after it dispenses ball number 10, an internal coin toss determines whether to add an additional 990 balls. They drop down from the soundproof reservoir, and the machine keeps on dispensing without missing a beat. It continues until it's empty.

The external behaviors of the Carter-Leslie and Eckhardt machines are identical. I am unable to distinguish them from the outside. Yet the inner workings make a difference. With the Carter-Leslie machine, the ultimate number of balls is "predetermined" before the first ball is dispensed. My random-ish selection is from the full group of balls. I'm much more likely to observe number 7 if there are only ten balls.

With the Eckhardt model, the ultimate number of balls may depend on a future event. Observing a low number like number 7 tells me that the crucial coin has yet to be tossed. Conditional on that, my quasi-random draw has been from the ten balls in the initial allotment. The chance of drawing number 7 is the same whether the machine is destined to have a total of ten or one thousand balls. No inference is possible.

With either machine, I might have observed a high number like 691. This would prove that the machine has one thousand balls. But if I observe a low number like 7, then I can't rule out ten balls or one thousand balls. I need to know what kind of machine it is to calculate the odds.

Now, all we have to do is decide which machine better corresponds to reality.

Anybody?

A person's existence in time and the end of the world are not determined by numbered balls. These are outcomes of countless interrelated and ongoing chance events. The flap of a butterfly's wings in ice-age Argentina may have determined the moment of my birth. A commuter train derailment in 1967 Baltimore could have predestined the robot apocalypse of AD 3024.

We live in a world of chaos, one whose details often aren't known well enough to permit deterministic predictions. That is cause to regard many events as random. Self-sampling offers a quick, simple way of reasoning about them.

Its essential premise is that *I can think of myself as a random draw from the whole group.* This fits some situations better than others. The Eckhardt token dispenser contains a booby trap nullifying this premise. Learning of a low number tells me that the coin toss that I'm trying to guess hasn't even happened yet—so I can forget about getting any clue from the number on the ball.

In order to reason from our metaphors, we have to agree on the details of the imagined random sampling. In their learned articles doomsayers and debunkers have been envisioning different sampling procedures. They didn't always spell this out, but it's implicit in their words and math. A seemingly minor detail can make a big difference.

"A Shooting-Room View of Doomsday"

Both the Carter-Leslie and Eckhardt models depart from reality in assuming a single, all-important coin toss. Actually our fates are determined by an endless succession of chance events. This is much better captured in the shooting room. In his 1997 paper, "A Shooting-Room View of Doomsday," Eckhardt finds Leslie's thought experiment to be the skeleton key to understanding the doomsday argument.

Eckhardt prefers a nonviolent version of Leslie's tale, one he calls the "Betting Crowd." A casino offers even-money, $100 bets on a roll of dice. As long as the dice come up anything except double sixes, the bettor doubles her money.

The deal is too good to pass up! Tour buses come from far and wide. Bettors wind around the block waiting to be admitted successively in groups of 1, 9, 90, 900, 9,000, 90,000... To keep the line moving, each group bets on the same roll of dice. After the roll they must leave to let others have a chance.

Unusual as Eckhardt's casino is, it shares a feature with those in Las Vegas, Monte Carlo, and Macao: the house always wins. Eventually double sixes come up, and everyone then in the casino loses. Given the admission system, 90 percent of all who enter the casino lose. Ninety percent of all bets lose. No ordinary casino can claim that kind of guaranteed profit.

We are left wondering how a bet can be incredibly favorable to the bettors *and* to the house. The answer, says Eckhardt, is that the casino is pulling one of the oldest tricks in gambling lore, "martingale." This is the dangerously ineffective gambling system in which a losing player doubles his bet, and keeps doubling it, until he wins.

Example. I bet $1. Lose. Bet $2. Lose. Bet $4. Lose. Bet $8. Win!

I've lost $1+$2+$4=$7. But I've won $8 for a net profit of $1. Martingale promises a gain of one betting unit for each series of bets ending in a win.

The system works, most of the time. But its fatal flaw is well known to all who gamble seriously. There's a small chance the bettor will suffer a long streak of losses, requiring him to wager more than he has. He will have no choice but to quit the game with a crushing loss. It is this small chance of ruin that balances the likelihood of a trivial gain.

Eckhardt's casino turns the tables by playing a martingale strategy on its customers (and investors). The risk is that the casino won't have enough money to cover the payouts for long runs of winning rolls for the customers, and/or that it won't be able to assemble the required crowds of bettors. Any casino that tried this would end up in bankruptcy court, with 90 percent of its customers getting pennies on the dollar of their winnings.

Let's grant suspension of disbelief as we did with Leslie's story. Say the casino has infinite money and an infinite crowd of eager bettors. As a bettor the only odds that matter are those of the dice. You should accept the casino's bets. Teams should organize to stand in line and place bets. So maybe a typical team of bettors wins thirty-five $100 bets and loses one. They've made a $3,400 profit. The casino can have a big sign outside saying, in gold letters, 90 PERCENT OF ALL WAGERS LOSE. It's true! But it doesn't matter. You'd be crazy *not* to put down your money.

As Eckhardt says, the bettor should tell himself: "Ninety

percent of all players will lose, but I have less than a 3 percent chance of belonging to that losing majority." These words have an "air of paradox," but the situation really isn't hard to understand.

The same reasoning can be transferred to the Carter-Leslie doomsday argument. Accept common assumptions about future population, and most people will live just before doomsday. This would be a simple demographic fact.

The unjustified step is translating that into a statement of probability that must apply to you or me. What's true of most people may not be true of *us*. We may know our particular situation well enough to disregard the overall numbers.

It is the premise that we can lay meaningful odds on doomsday that distinguishes the Carter-Leslie doomsday argument from Gott's Copernican method. We therefore need to consider exactly what these "prior probabilities of early doom" are. Here's one example. In 2003 Martin Rees created a stir by estimating the chance of civilization surviving the twenty-first century at only 50 percent. He based this estimate on his assessment of the risks of global war and nuclear, biological, and nanotechnological terrorism (*not* factoring in the doomsday argument). I hope Rees is wrong. But that's the sort of prior probability envisioned by the Carter-Leslie argument: a thoughtful synthesis of the risks confronting us, based on all available data.

Should we then use our position in time to adjust Rees's estimate, bringing doomsday even closer? No. Rees's pessimism most definitely comes with a time stamp. He was not saying we had only a 50 percent chance of surviving the eighteenth century. He was speaking of uniquely twenty-first-century risks, those tied to technology and population that are without precedent in human history.

This is not Rees's idiosyncratic belief. Anyone who thinks about it must agree that existential risk is not constant with time.

Caligula could not have pushed a button to kill everyone on Earth. When scholars or think tanks estimate existential risk, they are *of course* considering our moment in time.

It follows that I can't pick out a prior probability applicable to my epoch and then turn around and adjust that prior for my position in time. That would be double-counting evidence. Any well-informed assessment of existential risk already incorporates our position in time. That leaves little or nothing to be learned from doomsday reasoning.

"There may exist a plethora of reasons for supposing the human race to be doomed," Eckhardt wrote, "but our own birth rank in the total human population cannot reasonably be counted among them."

"The Doomsday argument does not fail for any *trivial* reason," Bostrom wrote. It has commanded extraordinary debate because, well, that's what philosophers do. Controversies get published, while consensus trivialities perish.

Doomsday is instructive: it shows self-samplers where some of the quicksand lies. One problem occurs when competing hypotheses predict greatly different numbers of observers. This can lead to disputes over SIA, which remains an area of controversy. Another, more general problem is when sampling procedures are left vague or unspecified—in particular, when there is no consensus about the appropriate reference class.

Fortunately not all applications of self-sampling raise such difficulties. Bostrom proposes a litmus test:

> Paradoxical applications are distinguished from the more scientific ones by the fact that the former work only for a rather special set of reference classes (which one may well reject) whereas the latter hold for a much wider range of reference classes (which arguably any reasonable person is required

> not to transgress).... I wish to suggest that insensitivity (within limits) to the choice of reference class is exactly what makes the applications... scientifically respectable. Such *robustness* is one hallmark of scientific objectivity.

Bostrom attempted to take this further. In his doctoral work he sought to lay out rules for choosing reference classes, rather than leaving this a matter of opinion. He succeeded in showing that extremely narrow or extremely wide reference classes can lead to absurdities and ought to be rejected. But Bostrom concedes that these guidelines are "quite weak."

Nearly twenty years later, he feels the reference class problem remains unsolved. "The last stretch of the PhD thesis was done in a bit of a rush," he said, "because I had applied to this postdoctoral fellowship at the British Academy and got short-listed. In order to accept it, I would have to have finished my PhD by a certain date." Thus his work ended abruptly. The dissertation closes on this wistful note: "I feel that the problem of the reference class... [may enclose] deep enigmas.... I hope that others will see more clearly than I have and will be able to advance further into this fascinating land of thought."

Part II

Life, Mind, Universe

Self-sampling can be applied to big questions of existence. The following chapters ask whether our world is a digital simulation; why we see no evidence of extraterrestrial intelligence; whether the origin of life on Earth was an unlikely accident; whether our universe is part of a multiverse. We take stock of the causes that might bring an early end to the human race, and we show why so many are concerned about artificial intelligence. In the final chapter, your author offers his opinions on doomsday and other matters.

The Simulation Hypothesis

Beginning in the 1920s, one of America's wealthiest men spent millions on a secretive project to counterfeit history. He was John D. Rockefeller Jr., son of the oil tycoon and one of the nation's most deep-pocketed philanthropists. "Junior's" intention was to develop Williamsburg, Virginia, as a living history museum. He bought up the town anonymously, lest property owners hear the Rockefeller name and raise their prices. The surviving colonial structures were then refurbished to simulate how they might have appeared when new, in the eighteenth century. Long-gone buildings were reconstructed. Where necessary, brand-new "colonial" ones were invented to create the illusion of a functioning colonial town.

Rockefeller envisioned a walk-through trip to America's past. Though deplored by some historians, that conception has been an enduringly popular tourist attraction. Most visitors arrive by car or tour bus, vehicles that must be parked in a lot and cannot be driven onto the town's quaint streets. Colonial Williamsburg's employees dress in eighteenth-century costumes and speak in a version of eighteenth-century vocabulary, grammar, and diction. Some play the roles of specific colonists, prominent and obscure, free and enslaved. They pretend not to notice visitors checking phones and jet planes passing overhead.

Colonial Williamsburg is "accurate-ish," one executive admitted. It demonstrates that it is not just the momentous events of history that capture our imagination but also the mundane details—even if some of them are fudged. Renaissance fairs, Civil War reenactments, TV and movie costume dramas, and history-themed video games also offer ways to reconnect with the past. Unless human nature changes fundamentally, it's hard to conceive of a future in which *nobody* is curious about the past.

This leads us to the *simulation hypothesis,* the claim that the world we experience is an artificial, digital simulation, an immersive "video game" created by a technologically advanced society. It's a familiar trope of science fiction, and lately some well-informed people are taking it seriously.

In 2016 a panel of scientific luminaries met at the American Museum of Natural History for a debate on the subject. Moderator Neil deGrasse Tyson offered 50:50 odds on the simulation hypothesis being true. "We would be drooling, blithering idiots" compared to the humans of the future, Tyson said. "If that's the case, it is easy for me to imagine that everything in our lives is just a creation of some other entity for their entertainment."

"If I were a character in a computer game," MIT physicist Max Tegmark said, "I would also discover eventually that the rules seemed completely rigid and mathematical. That just reflects the computer code in which it was written." He likened this to the mathematical nature of physics.

Harvard physicist Lisa Randall disagreed. She put the probability of our being a simulation at "effectively zero." For her the real question was "why so many people think it's an interesting question."

One who takes simulations seriously is entrepreneur Elon

Musk, who has helped fund Nick Bostrom's work. "The strongest argument for us being in a simulation," Musk said at the 2016 Recode conference, "is the following: 40 years ago, we had Pong. Two rectangles and a dot. Now 40 years later we have photorealistic 3D simulations with millions playing simultaneously. If you assume any rate of improvement at all, then the games will become indistinguishable from reality. It would seem to follow that the odds that we're in base reality is 1 in millions."

In 2016 a *New Yorker* profile of venture capitalist Sam Altman mentioned in passing that "two tech billionaires have gone so far as to secretly engage scientists to work on breaking us out of the simulation." This promptly led to speculation that one of the billionaires was Musk. Others wondered how it was even possible for simulated beings to break out of their simulation. Journalist Sam Kriss complained that "the tech industry is moving into territory once cordoned off for the occult." *New York Times* science writer John Markoff called the simulation hypothesis "basically a religious belief system in the Valley"—Silicon Valley, naturally.

The Omphalos Scenario

How has the simulation hypothesis gained such intellectual currency? Does it merit being taken even a *little* seriously? The answer has to do with the self-sampling assumption, and with a set of beliefs ingrained in contemporary culture.

The notion that the world *could* be an illusion is as old as philosophy. Plato's cave, and all that. It was, however, an eminent Victorian, Philip Henry Gosse (1810–1888), who took Plato's idea to the next level. Gosse was a naturalist who invented the aquarium and corresponded with Darwin on orchids. His 1857 book, *Omphalos: An Attempt to Untie the Geological Knot,* considers this puzzle: Did Adam and Eve have navels? (*Omphalos* is Greek for "navel.") A navel is the scar of the umbilical cord. Scripture is clear that Adam was not born of woman, and Eve was created from Adam's rib.

Gosse maintained that the first couple did have navels. God supplied the illusion of a past-that-never-was, for the sake of a harmonious creation. Eden's trees would have had growth rings.

Gosse further believed that God created fossils, although the creatures represented had never existed. He even argued for the divine creation of coprolites (the distinguished term for fossilized excrement). Recognizable by their shape, coprolites are the most common type of mammalian fossil. Admitting that coprolites were "considered a more than ordinarily triumphant proof of real pre-existence," Gosse claimed that they too were part of the meticulous stage-set illusion of God's creation. "It may seem slightly ridiculous," conceded Gosse, "but truth is truth."

The faithful did not see this as truth at all. Even less did *Omphalos* hold sway in the Darwin-disrupted scientific world. Yet Gosse's deeply misguided book, too nutty to forget entirely, has echoed down through the years. In the twentieth century Bertrand Russell distilled the omphalos scenario into this philosophical riddle: Suppose the universe was created five minutes ago. How would you know otherwise?

The knee-jerk reaction is, we've all got memories going back more than five minutes. We've got documents backing up those memories. *All fake!* said Russell. Maybe you and your memories came into existence five minutes ago. Same for Stonehenge and T. rex fossils. The Berlin Wall and the Twin Towers never existed—they're just memories implanted in our minds.

Unlike Gosse, Russell wasn't saying this was true. He was just making the point that there are some things we can never know with certainty.

The modern simulation hypothesis is generally traced to computer scientist–entrepreneur Stephen Wolfram and his 2002 book, *A New Kind of Science*. Wolfram made the case that our world might literally be a digital simulation. He presented this idea as a testable hypothesis. It might be possible to look for evidence of "pixelization" in subatomic physics. But not a few reviewers of Wolfram's book pegged him as a genius who had gone off the deep end.

Bostrom's Trilemma

In 2003 Nick Bostrom took up the theme. It is Bostrom's elaboration, framed as an application of self-sampling, that has gained so many serious and semiserious converts.

Bostrom is not saying we are simulations. Neither is he simply making a philosophic point about what could be true "for all we know." He is saying, rather, that the simulation hypothesis is several degrees less outlandish than it first appears. It is a case where it is difficult to assign a probability.

The core of Bostrom's idea is *ancestor simulations*. Advanced societies may be able and willing to create all-encompassing digital simulations of the past. They will not be the half-convincing imitations of today's theme parks or virtual reality but rather something totally seamless, indistinguishable from the real thing. Since the point would be to create a faithful historical reconstruction, most of the simulated people would not be permitted to know that they are simulations. Such knowledge might change their behavior, breaking the fourth wall. Thus simulated worlds wouldn't be expected to have signs saying DISCLAIMER: THIS IS ONLY A SIMULATION.

Now apply the self-sampling assumption. By definition there can be only one real world. There could be many, many simulated worlds (read on to see why some believe this). If so, then simulated observers outnumber real ones. I can't know which kind of observer I am—since the simulations are indistinguishable from the real. So, as a random observer, the odds favor my being a simulation.

The commonsense objection is that simulation technology hasn't even been invented yet. Well, we don't know that. Grant even a sliver of credibility to Bostrom's idea, and our present could be a future society's simulation of its past. We *think* it's the twenty-first century because our calendars and phones tell us so, because history books go up to the early twenty-first century and stop, and because the History Channel never runs documentaries on interstellar wars of the thirty-second century. Maybe that's all part of the simulation—a future society's homage to the long-ago twenty-first century.

Bostrom isn't saying this is necessarily true. He's just describing the conditions for it to be a valid conclusion of self-sampling. Bostrom asserts that at least one of three claims has to be true:

1. Technology capable of creating ancestor simulations will never exist.
2. Nobody will ever want to create ancestor simulations (even if they have the ability).
3. We're probably living in a simulation right now.

This is known as *Bostrom's trilemma*. Its logic is as simple as its conclusion is jaw-dropping. Either number one or number two has to be false; otherwise, number three must be true. The sci-fi cliché, of the android that doesn't know it's an android, could be us.

Are Simulations Even Possible?

Maybe we're getting *way* ahead of ourselves. Why should we believe that seamless world simulations, indistinguishable from reality, are possible?

Any technophile will agree with Musk that our AV gear keeps getting more awesome. Virtual reality will doubtless shed today's unconvincing textures, stereotyped facial expressions, and seasick-inducing lags in response to achieve perfect visual and auditory fidelity. This might be the easiest part of Bostrom's idea to accept.

The system would have to cover all the other senses too, but okay, that sounds doable.

The simulation of an entire world of people is something else again. The required computing power would be incredible. Bostrom offers back-of-the-envelope calculations indicating that world simulation is, well, not impossible. But it might require planet-sized computers.

An ancestor simulation would not just be a hyper-detailed video game. It would also incorporate the virtual people experiencing that world-game. It would have to simulate human minds in full detail, requiring full-fledged artificial intelligence.

It has been estimated that the neural firings of a human brain are equivalent to about 10^{16} to 10^{17} processor operations per second. Right now there are supercomputers that do 10^{17} operations a second. If only we knew how the brain works, we'd be good to go. We could simulate a single brain in real time.

But an ancestor simulation would encompass the stream of consciousness of everyone alive at the point in time being simulated. At the present world population this would require something like 10^{33} to 10^{36} operations per second.

A comprehensive simulation would also need to incorporate buildings, cities, roads, forests, deserts, oceans, weather, and sky. This might sound overwhelming. Actually it's probably less of a challenge than the simulation of minds. There would be little point in simulating every atom, chloroplast, or gnat. The simulation's detail could be highly selective, favoring those parts of the environment that people notice. The Earth's molten core plays no direct role in human affairs. Thus the zone of simulation might extend only a few feet into the soil. (Dig a very deep hole, and the simulation would invent the soil beneath the blade of the shovel.)

We are more affected by floods, blizzards, hurricanes, earthquakes, and volcanic eruptions. Chaos theory says these phenomena defy prediction. Did it rain on Charles and Di's wedding? (No. It was sunny with scattered clouds.) The simulation could draw on weather reports and news accounts to supply historically accurate weather and catastrophes.

The simulation's sun, stars, and planets could in effect be planetarium projections. Further detail could be generated only as needed. The Apollo moon landings might require simulating a few acres of lunar surface. Whenever a simulated biochemist sequenced a genome, or a sim physicist ran an accelerator experiment, the code would invent a level of detail otherwise lacking.

Bostrom estimates that a brute-force world simulation, say of Earth with its twenty-first-century population, might require 10^{33} to 10^{36} operations per second. In comparison Bostrom estimates that a planet-sized computer might be capable of 10^{42} operations per second.

Planet-sized computer? Freeman Dyson, and transhumanists generally, contemplate a future in which humanity might be able to marshal much of the mass and energy of a planet, a star, or many star systems. From that lofty standpoint, ancestor simulations might be feasible.

Bostrom certainly isn't ruling it out. He figures that "a [planet-mass] computer could simulate the entire mental history of humankind... by using less than one millionth of its processing power for one second. A posthuman civilization may eventually build an astronomical number of such computers."

Colonial Williamsburg has an actor playing Edith Cumbo. We know Cumbo's name, and that she was a free black woman, born about 1735, who headed her own household in Williamsburg. No one knows her occupation, what she looked like, or when she died. Cumbo's name appears in various legal documents, none of them jelling into much of a biography. On June 15, 1778, Cumbo sued a certain Adam White for trespass, assault, and battery.

The future will know far more about us than we know about our ancestors. Starting in the early twenty-first century, ordinary people began documenting their lives on social media. Our Facebook and Instagram feeds would be helpful to simulators. Is it just a coincidence that we find ourselves in the social media age?

DNA tests are becoming cheap and popular. The companies offering them swear the results are private. Yet data, once it exists, has a way of turning up and being put to unexpected uses. The *Bodleian Plate,* an eighteenth-century engraving of Williamsburg discovered in 1929, guided Rockefeller's reconstruction of the town.

Bostrom's conception of world simulations assumes the development of artificial intelligence that can pass a robust Turing test and behave as a psychologically convincing human. Wrap that code in an avatar, and you've got a virtual human. A World War II simulation could include representations of Churchill, Hitler, and Roosevelt, embodying everything known about these people. More than that, the simulation could include battles, bond drives, fascist rallies, and USO shows in which every person is a psychologically realized simulation, supplied with name, rank, and serial number taken from military records, and any other information that may survive. Where information is lacking it could be invented to create realistically diverse crowds rather than an army of clones.

One reaction is that this would be an incredible waste of resources. It would, by today's standards. But when technology becomes quick, easy, and cheap, we find new uses for it. Those uses are wasteful and self-indulgent (say the older generation). The phone in my pocket is a more powerful computer than those that sent Neil Armstrong to the moon. Most of the time I use it for completely frivolous things.

There could be serious uses for ancestor simulations. Historians might want to explore simulated pasts for research. What would have happened if Truman had not dropped the atomic bombs on Japan? Thousands of simulations could reveal how large and small changes to initial conditions lead to different outcomes. History could become an empirical science. If some leaders were willing to learn from history, that could have immeasurable benefits.

Simulations might become so inexpensive and routine that kids are assigned simulation experiments for history class. They could affect travel. Would you rather vacation in modern Tuscany, with its museums, or in Renaissance Tuscany, meeting virtual Leonardo, the Medicis, and Machiavelli?

Tourists, gamers, genealogy buffs, historical cosplayers, experimental historians... There would be a demand for ancestor simulations, and the simulations would outnumber the unique reality. The demographics of Williamsburg provide some support. The town and its surrounding county had a population of about 5,000 at its eighteenth-century peak of influence. Today's Colonial Williamsburg draws about 480,000 visitors a year. The great majority of people who have experienced Williamsburg have done so not as eighteenth-century settlers but as tourists from the future.

The Ethics of Simulations

A world simulation might include a few tourists from the future, inhabiting a particular avatar in order to mingle with the crowds, unrecognized. Those crowds would mainly be autonomous, artificially intelligent virtual beings. They would talk, act, and react just like regular humans. This raises an important question: Can the simulated people experience consciousness?

The answer would have to be yes for the simulation hypothesis to be true. Otherwise the fact that you are now experiencing consciousness (I assume you are....) would prove to you that you're *not* a simulation. If simulations are mindless zombies, then they can no more be in your reference class than a head of lettuce is.

Most of today's AI researchers, and most in the tech community generally, believe that something that acts like a human and talks like a human and thinks like a human—to a sufficiently subtle degree—would have "a mind in exactly the same sense human beings have minds," in philosopher John Searle's words. This view is known as "strong AI."

Searle is among a dissenting faction of philosophers, and regular folk, who are not so sure about that. Almost all contemporary philosophers agree in principle that code could pass the Turing test, that it could be programmed to insist on having private moods and emotions, and that it could narrate a stream of consciousness as convincing as any human's. But this might be all on the surface. Inside, the AI-bot could be empty, what philosophers call a zombie. It would have no soul, no subjectivity, no inner spark of whatever it is that makes us what we are.

Bostrom's trilemma takes strong AI as a given. Maybe it should be called a quadrilemma, with strong AI as the fourth leg of the stool. But for most of those following what Bostrom is saying, strong AI is taken for granted.

If simulated people have real feelings, then simulation is an ethically fraught enterprise. A simulation of global history would recreate famine, plague, natural disasters, murders, wars, slavery, and genocide. This would mandate billions of virtual deaths of virtual beings who experience pain and despair. That would make the simulators as bad as every villain of history rolled into one.

Then there is "Roko's Basilisk," an urban legend of the transhumanist community. The Basilisk is a morally defective future AI that blackmails people to do its bidding. Do as the Basilisk wants, and nobody gets hurt. Otherwise it simulates many exact copies of you in very unpleasant worlds. Note that the Basilisk may already exist, for its "future" could be our "now." You have to know about the Basilisk, and think about it, for its threats to have any power over you. So maybe I've already said too much. Forget you ever read this paragraph, okay?

That simulations have feelings bears on Bostrom's claim number two. Societies capable of ancestor simulations might ban them for ethical reasons. Or else they'll restrict simulations to alternate realities in which only nice things happen. We're not living in that kind of simulation.

Maybe it's possible to create functional simulated beings with or without consciousness, as desired. Ethical simulators would create zombie sims only to populate their ancestor simulations, and the simulation argument need not apply—for anyone who experiences true consciousness.

Another thought: outlaw simulations, and only outlaws will have simulations. That ups the odds we're living in the simulation of a mad scientist, psychopath, or Basilisk.

Why I Might Be Real

The simulation hypothesis has become reliable clickbait. The media often fail to distinguish Bostrom's nuanced trilemma from "Scientist Says We're Living in the Matrix." Bostrom's contribution is in laying out what you'd have to believe in order to conclude that we are likely to be living in a simulation. To most in tech culture, it is as natural as breathing to accept that computing power will grow indefinitely, and that every killer app conceived will be used (even if the luddites don't like it). To those outside the silicon bubble, these claims may be less convincing. The simulation hypothesis offers an object lesson in how people in tech think differently.

Yet it is possible to accept the technologically audacious premises and use them to make a Bayesian case for probably *not* being a sim. Start with the reference class. If the consciousness of real-me is truly indistinguishable from that of sim-me, then we both must be in the same reference class. This is one of Bostrom's "quite weak" restrictions on reference class, and it is generally accepted.

The tricky part is identifying one's self-locating information. Admit that the simulation hypothesis might be true, and I can no longer reason from the apparent circumstance that I live in the base reality of twenty-first-century Earth. I could be code in a planet-sized computer circling Betelgeuse in AD 35,000.

What I know is this, that—*taking things at face value*—I live in a world where simulation technology has not been invented. That's what I've got to work with, and even that qualified statement has Bayesian implications.

- If sim technology does not now and never will exist, then I am 100 percent certain to find myself in a world without sim technology.
- But if sim technology exists or is destined to exist, then the chance of finding myself in a world apparently without sim technology is something less than 100 percent. It's less because there has to be a group of real people who created the simulations, a world where sim technology is accepted as an everyday fact. I'm not in that group or that world.

My personal situation is more probable with the first hypothesis. It's not clear by how much. If real people are a tiny, tiny minority relative to the sims, then even the second hypothesis might result in a high probability for being in a world without evident sim technology. There would not be much Bayesian leverage to favor either possibility.

It looks like sims would require cosmic-scale engineering. You don't turn a planet into a gaming console unless you've got a lot of planets to play around with. Any society capable of creating world simulations would have mastered space travel. It would have spread to other planets and star systems over thousands of years, achieving immense cumulative populations. The population living after the invention of sim technology could be much, much greater than the "current" population.

Simulation technology would be a transformative thing, like television or the internet. The real people in a society possessing, and benefiting from, simulation technology would be well aware of it. And if the technology had existed for a long time, more and more ancestor simulations would revisit epochs *after* sim technology was invented. There would be sims of people who knew they lived in a society with omnipresent sim technology. There would be sims of sims.

It's hard to escape the conclusion that sim-creating societies would accept the simulation hypothesis as a fact of life. Children would learn Bostrom's trilemma along with their ABCs. Every four-year-old would know she *could* be a sim, and probably *is* a sim. Dad: "It's nothing to worry about." Mom: "It's *normal.*"

The diagram below divides human-like consciousness into four categories. There are real minds and simulated minds, minds living before the invention of sim technology and after it. The areas of the four rectangles indicate the relative populations schematically, as measured by head count or observer-moments. The population living after the invention of sim technology is much greater, as that supposes that humans have spread to many other planets.

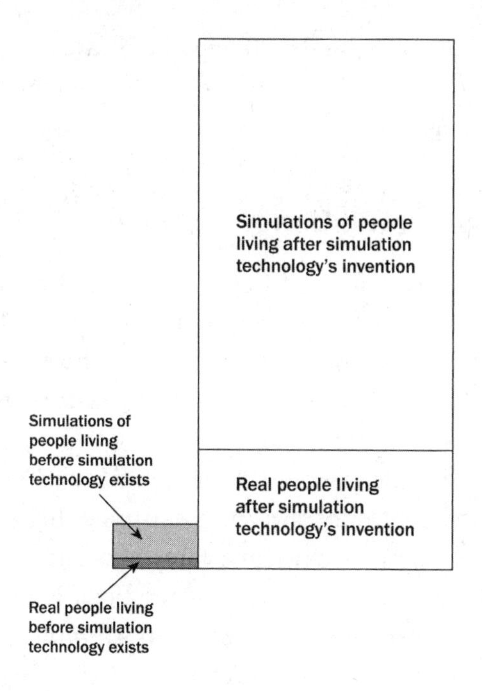

Because I live in a society that does not seem to have sim technology, I must be in the shaded area of the diagram. I don't know whether I'm a real person (dark shading) or a sim of some guy who lived prior to sim technology's invention (light shading). Either way, I'm unusually early in the timeline of simulation technology. This is highly improbable if sims exist or will exist. It is certain if there are not and never will be sims. That is Bayesian cause for believing that there will be no sims—and I am not a sim.

This analysis mimics the doomsday argument. As with doomsday we ought to ask whether self-sampling is the best tool for the job. Can I set a probability of being a sim, perhaps one tailored to my precise moment in time? This might be much more relevant than the estimate provided by self-sampling. But it's hard to see how anyone can have a strong, informed opinion about whether they're fake. Also, the simulation hypothesis says that I don't even know what my moment in time is. Here self-sampling appears to be the viable approach.

Testing the Simulation Hypothesis

In the absence of data, you should go out and get some, Elliott Sober said. Proposals for testing the simulation hypothesis occasionally turn up in scientific journals. As Wolfram suggested, we should be looking for "jaggies," for telltale signatures of pixilation in physics.

In 2012 Silas Beane and colleagues identified one potential signature. They concluded that the spectrum of high-energy cosmic rays might show evidence of a computed, lattice structure to physics. The accuracy of current observations is insufficient to make the distinction, but they are within a few orders of magnitude of being so. A test of the simulation hypothesis might be possible in the near future.

But, like everything else attached to this idea, it depends on a network of untestable assumptions. One is that the supposed simulation will use voxels (the 3-D equivalent of pixels) to represent space. This is the way movie digital effects and implementations of virtual reality work. It's anyone's guess whether posthuman world simulators would use that design. They might have something better.

Another assumption is that the simulators aren't trying too hard to keep us from learning the truth. Bostrom suggests that the simulation software might constantly keep tabs on what its simulated minds are thinking and doing. Every time a sim tries to do something that would expose the simulation, countermeasures could be taken. If Bostrom is correct, then the directors of our simulation are already aware of Beane's cosmic rays test. Were it to be implemented, the simulation could generate higher-resolution detail, nullifying the looked-for signatures. "Should any error occur," Bostrom writes, "the director could easily edit the states of any brains that have become aware of an anomaly before it spoils the simulation. Alternatively, the director could skip back a few seconds and rerun the simulation in a way that avoids the problem."

Learning to Love the Matrix

The simulation hypothesis and the doomsday argument tend to be mutually exclusive. Should we be living in a simulation, it would mean our species survived all the crises that now concern us. We didn't blow ourselves up; CO_2 didn't turn the planet into another Venus; the robots didn't turn against us.

If, on the other hand, the end is near, then we'll never create those world simulations. Our reality is all there is, and that makes our headlong rush to catastrophe all the worse.

Would you rather be a mere simulation of a successful species or a flesh-and-blood member of a doomed one? Put that way, maybe it's possible to learn to stop worrying and love the Matrix. Being a sim won't affect what you'll eat for dinner, whether you'll get that promotion, or where you'll go for vacation next winter. Unless someone pulls the plug before then. As Robin Hanson put it, "Your motivation to save for retirement, or to help the poor in Ethiopia, might be muted by realizing that in your simulation, you will never retire and there is no Ethiopia."

A disconcerting thought is that our simulation could be like Russell's omphalos scenario. Maybe future historians are interested in President Donald Trump because of some momentous effect he had on subsequent world history. In order to understand the dynamics of the Trump era, the historians need to play a certain five-minute segment of it over and over with slightly different initial conditions. Our world is one of those five-minute loops, a *Groundhog Day* moment for a future historian's PhD thesis.

The Fermi Question

Where is everybody?"

Physicist Enrico Fermi asked that question, to a burst of laughter, one radiant summer day in Los Alamos, New Mexico. He meant extraterrestrials. Why aren't intelligent beings from other planets visiting the Earth in spaceships?

It was 1950, and Fermi was having lunch at Fuller Lodge, Los Alamos National Laboratory. He was half-joking about something in *The New Yorker,* a cartoon playing off news reports of "flying saucers." He was also half-serious. Fermi had done the math. There are billions of stars in our galaxy. Many must have Earth-like planets. Some of those planets should have developed intelligent life before Earth did and possess technology far beyond our own. Fermi suspected that advanced civilizations would have ways of traveling faster than the speed of light.

"How probable is it that within the next ten years we shall have clear evidence of a material object moving faster than light?" he asked.

The question was put to physicist Edward Teller, whose answer was 1 in a million.

"That is much too low," Fermi said. "The probability is more like ten percent." (The media called Fermi the "architect of the atomic bomb" and Teller the "father of the hydrogen bomb." Fermi had created the first nuclear chain reaction in 1942, on a squash court at the University of Chicago. The site is now a tennis court. Fermi was an avid and aggressive tennis player.)

Fermi was inclined to believe that faster-than-light travel would have already been discovered by extraterrestrials. Uncounted races ought to have explored the whole galaxy, including the Earth. Hence Fermi's question, which came as a delightful lunch-break non sequitur. One witness recalled, "In the middle of this conversation, Fermi came out with this quite unexpected question, 'Where is everybody?'... The result... was general laughter because of the strange fact that in spite of Fermi's question coming from the clear blue, everybody around the table seemed to understand at once that he was talking about extraterrestrial life."

The Fermi question—*Where is everybody?*—is rhetorical. It has also been described as a paradox, for the obvious answer—that intelligent life is much rarer than we think—is one that many find difficult to accept. The Los Alamos setting, with its cast of fission and fusion bomb patriarchs, invited one explanation. Maybe technological species are rare because they don't last very long. They annihilate themselves in global war before they get around to exploring the galaxy.

"What we all fervently hope," Fermi once said, "is that man will soon grow sufficiently adult to make good use of the powers that he acquires over nature." Privately Fermi believed atomic weapons would lead to war. His Manhattan Project colleague, mathematician John von Neumann, minced no words. He rated it "absolutely certain (1) that there would be a nuclear war; and (2) that everyone would die in it."

Drake Equation

The consensus of biologists and screenwriters is that we are not alone in the universe. This is not a new idea. Dominican friar Giordano Bruno, a supporter of Copernicus, asserted that stars are suns, circled by planets harboring intelligent beings. Pope Clement VIII had him burned at the stake in 1600. The church fathers considered Bruno's doctrines so heretical that he was not permitted any last words. His tongue was nailed to his mouth with iron spikes.

By the early twentieth century, many accepted the possibility of intelligent life on Mars. Guglielmo Marconi hoped that the new technology of radio would permit contact with Martians. Listening on his yacht *Electra,* in the Mediterranean, Marconi picked up signals from the Red Planet in 1919. Or thought he did. In 1922 Mars made a close approach to Earth. The world's radio stations politely observed silence to help Marconi and others pick up signals. No one heard anything convincing.

In the spring of 1960 American radio astronomer Frank Drake made another attempt to detect signals from extraterrestrial intelligences. Drake had far more sophisticated technology and no illusions about life on Mars. He directed West Virginia's eighty-five-foot-diameter Green Bank radio telescope to two nearby sun-like stars, Tau Ceti and Epsilon Eridani. The results were negative, but media coverage of the attempt captured the public imagination.

The following year Drake convened a meeting of interested scientists at Green Bank. He asked his visitors to consider how many intelligent species exist in our galaxy. That number, Drake said, is the product of seven unknowns:

(1) how many stars come into existence in our galaxy per year;
(2) how many of those stars have planets;

(3) how many planets exist in a typical star system;
(4) how many such planets develop life;
(5) how many of those life-bearing planets evolve intelligent life;
(6) how many intelligent species broadcast radio signals (or otherwise reveal their existence);
(7) the lifetime of communicating intelligent species.

Drake's thinking was thoroughly Copernican. The Earth, its sun, and *Homo sapiens* are typical until proven otherwise. You can divide the Drake factors into two groups. Factors (1) through (3) are matters of astronomy, with grounding in data. But (4) through (7) involve speculations about extraterrestrial biology, history, and motivations. Factor (7) in particular was recognized as a wild card. The span of time in which ETs are willing and able to broadcast signals affects the prospects for detecting a signal.

The group decided that the chance of a suitable planet developing life (4) was essentially 100 percent. They gave the same optimistic value for the chance of intelligence arising (5). It was estimated, in fact, that the product of all the first six factors was, very approximately, 1. Thus it came down to the last factor, the civilization lifespan. On this point opinions ranged widely, from 1,000 to 100 million years. That in turn led to the final estimate of 1,000 to 100 million presently communicating civilizations in our galaxy.

An estimate multiplying seven unknowns can hardly be expected to be accurate. The conference scientists were not unaware of that, nor of being a self-selected group interested in extraterrestrial life. It's possible their numbers were skewed in favor of many ET species. But the bigness of the universe seemed to win out. There ought to be *lots* of ETs out there.

Since Drake devised his formula, much has changed. More than 3,800 planets orbiting nearby stars have been discovered.

This provides compelling evidence that practically all sun-like stars have planets (factor 2) and that there are almost always multiple planets in a star system (3). In those regards, the Green Bank estimates were conservative. But now as then, much of the uncertainty in the Drake estimate comes from that last doomsday factor. How long do extraterrestrial civilizations survive?

Von Neumann Probes

"Absence of evidence is not evidence of absence," said Martin Rees, Cambridge cosmologist and Astronomer Royal. Just because we don't have evidence for ETs doesn't mean they don't exist.

Astronomer and science fiction author Glen David Brin termed that lack of evidence the "Great Silence." Over the years many explanations have been put forth. It is not hard to come up with believable rationalizations for why *some* ETs might not want to communicate with us, or why they might not be interested in exploring ("colonizing") space, or why they might meet an early end. The challenge is to come up with an explanation that would apply to practically *all* ETs. Should only a tenth of 1 percent of ET species be noisy gadabouts, there could still be many such in the galaxy, and the Fermi question would remain.

It's conceivable that interstellar travel is simply a nutty idea that "belongs back where it came from, on the cereal box." That was physicist Edward Purcell's 1960 assessment, at the dawn of the space age. The "island model" holds that the speed of light is a universal barrier, confining intelligent species to their home solar systems. Widespread galactic exploration, or even communication by radio waves, is discouragingly slow and just not worth it. That then would be the answer to the Fermi question.

In 1975 MIT astronomer John Ball advanced the "zoo hypothesis." He suggested that extraterrestrial explorers make a conscious effort *not* to leave traces of themselves. They could regard the galaxy as a national park or zoo to be left as pristine as they found it. And ETs may have no interest in communicating with the inhabitants of their zoo. (We have little interest in communicating with ibexes, much less in establishing diplomatic relations with every ibex herd or zoo specimen.) It could be that contact with a more advanced civilization is known to be devastating to the less advanced civilization. ETs might be avoiding us for our own protection.

Even in Fermi's time, there was a comeback for these ideas: von Neumann probes, described by John von Neumann. The best-known examples, fictional of course, are the black monoliths in Stanley Kubrick's *2001: A Space Odyssey.* An early cut of the film had a segment explaining exactly what the monoliths were. They were identified as self-reproducing machines, designed to explore space. Kubrick decided to cut the exposition, leaving the monoliths enigmatic and symbolic.

A von Neumann probe is a robot that is able to build copies of itself. (Who knows what it would look like? A rectangular monolith is as good a guess as any.) Such robots could be sent out to explore the galaxy in the stead of biological beings. Like Google Street View vans, they might systematically explore even the more uninteresting parts of the galaxy for the sake of a comprehensive survey. Some probes would become damaged or destroyed, but the others would be able to replace them by collecting raw materials and building new probes. This redundancy would give the collective mission a high chance of success.

Von Neumann felt that this would be the practical means of deep space exploration, given that sub-light-speed trips to the stars would exceed human lifespans. The probes could do anything an inquisitive human explorer might do, and they could transmit their findings back to Earth.

In recent years von Neumann's idea and elaborations of it (by Ronald N. Bracewell and Frank Tipler, among others) have entered the Fermi paradox conversation. As computer and robotic technology advances, von Neumann's idea is not so outlandish as it once seemed. It's been estimated that von Neumann probes could explore the whole galaxy, at well under light speed, in as little as a million years. It wouldn't even matter whether the species that created the first probes was still around. The robots could continue their mission regardless of whether there was anyone at home to receive their dispatches. Many now believe this idea is as credible as or more credible than ETs visiting Earth in spaceships. So... where is everybody/thing?

We might expect to find ancient von Neumann machines on Earth. They'd be "dead" or "sleeping" or fossilized. In some versions of the concept, the machines limit their numbers. In others they multiply like lemmings, as long as there are resources. In the latter case we might expect to find whole geological strata of von Neumann machines, documenting an extravagant binge in which they swarmed the globe, got whatever they wanted, and left.

But no one has ever found a convincing alien artifact of any kind. Fermi's question remains as great a mystery as ever. Nick Bostrom painted this word picture: life on Earth is a single data point, and the Fermi paradox is the question mark over it.

The Princess in the Tower

In 1971 Byurakan, Armenia, hosted a conference on extraterrestrial life that is remembered for a confrontation between celebrity astronomer Carl Sagan and biochemist Francis Crick, codiscoverer of the double helix. Their dispute was over a simple question: How likely had it been for intelligent life to arise on Earth?

The absence of evidence for ETs had only made Sagan's heart grow fonder. He was building a career as a cheerleader for the existence of extraterrestrial life. Sagan's position was that the Earth is a typical planet, as far as we can tell. Intelligent life evolved here. That's a one-for-one batting average for intelligent life.

Sagan further noted that the earliest then-known fossils of cyanobacteria dated back more than 3 billion years, to within the first third of the Earth's existence. "This, to me, speaks rather persuasively for a rapid origin of life on the primitive Earth," he said. Sagan held that an early start for life on Earth bolstered the case that life, and even intelligence, are not-improbable outcomes for Earth-like planets.

Crick disagreed. He summarized his position this way:

In order to display the difference between my position and Professor Sagan's, I have to make an analogy and I am sorry it is so conventional. Consider a man who has been dealt a hand of playing cards. The character of his hand is that he has to have one particular sequence, one particular combination of cards. We know that this is a rare event and it is not reasonable to try to estimate the probability of the event simply because it has happened to us. Professor Sagan's argument is that there are plenty of playing cards. But we have only a unique event and strict probability theory says that we are now allowed to deduce probability in that way.

Crick said, "I do not know whether such a line of reasoning has a name, but it might be called the 'statistical fallacy.'"

We have a better vocabulary for talking about it now. The term Crick was searching for is an "observation selection effect." We need to consider how we came by the evidence we have. It may be biased and not a random draw.

Whether intelligent observers are common or rare, we exist. We should not be surprised to find ourselves on a planet with sentient beings. And since it took quite a bit of evolving to get from earliest life to *Homo sapiens,* and evolving doesn't happen overnight, then we'd expect to find that life originated a long, long time ago, early in our planet's history. That is what we do find. Our existence tells us virtually *nothing* about the probability of life or intelligent life arising.

Crick's point is now generally accepted. It seems to present a stalemate. We can't learn anything about the probability of life from the mere fact of our own existence. But Sagan believed that the timing of life and intelligence on Earth offers additional evidence. This idea has since been explored in more detail. Brandon Carter illustrated it with a fairy tale (not Sleeping Beauty, but please bear with me for another somewhat dated gender role).

Lucky Chuck

A wise princess is locked in a tower, awaiting a worthy suitor. Anyone wanting to marry her must pick the lock on the tower's door in one hour's time. Should the suitor succeed, the marriage takes place immediately. Suitors who fail are beheaded.

The design of the tower's lock makes it impossible to run through combinations methodically. Suitors must try combinations at random.

The princess is so wealthy that there is always a long line of applicants, winding past the ever-expanding graveyard of failed suitors. One June day a lucky suitor named Chuck succeeds. He picks the lock in 27 minutes and 14 seconds.

Was Chuck's task easy or hard? How probable was it that Chuck succeeded?

This (says Carter) is not too different from the situation in which we find ourselves, asking how likely intelligent life is.

There are at least two ways to answer the question about the lock. The better, more direct way is to count the tombstones in the graveyard of failed suitors. Add one for lucky Chuck. We might guess the chance of success is something like one divided by that number.

But suppose the graveyard is hidden far away in the forest. Chuck does not know whether he was the first suitor or the ten thousandth. All he knows is that he succeeded, and that it took 27:14 out of the allotted hour.

He has no idea how long it ought to have taken, on average, to pick the lock. It could be ten seconds, ten days, or ten centuries. (The princess may believe that no one in the kingdom is worthy of her.)

All Chuck knows is that he could not have succeeded by taking more than an hour, for the clock would have run out, and the executioner awaited.

Given that he survived, Chuck's time to pick the lock had to be in the range of zero to sixty minutes. I'll break the possibilities into three cases:

(1) The average time to pick the lock is much less than an hour. The first suitor usually succeeds, and the time taken is generally a small fraction of an hour.

(2) The average time to pick the lock is on the order of an hour. Suitors have a decent shot at success, and also a substantial risk of failure. The time taken by a successful suitor could be any fraction of the hour.

(3) The average time to pick the lock is much, much more than an hour. Most suitors fail. When and if one succeeds, the time is any fraction of an hour.

You might question that last claim. Given that the lock is difficult, isn't it easier to believe that a suitor would succeed in the nick of time (say at 59:42) rather than right off the bat? In a movie the hero always cuts the wire on the time bomb with about two seconds to spare.

In reality, a nick-of-time success is not more likely in case (3). Carter specified that combinations must be tried at random. In effect, suitors are playing a slot machine. Every pull of the handle gives the same small chance of success, for the machine has no memory. The player is just as likely to succeed on the first attempt as on the millionth. Should the player hit the jackpot within the first hour of play, it is just as likely to occur at the beginning of the hour as at the end.

That means that Chuck's lock-picking time of 27:14 is consistent with cases (2) and (3). It fits in less well with possibility (1). The fact that it took Chuck twenty-seven minutes (rather than twenty-seven seconds) should tell him that it's probably not all that easy to pick the lock. But that's about the limit of what Chuck can conclude.

The evolution of intelligent life on Earth was also working against a time limit. This leads to three hypotheses, paralleling those above:

(1) The origin of intelligence is a slam dunk. Essentially all Earth-like planets evolve observers, and it usually happens early in the planet's habitable lifespan. (The "habitable lifespan" takes into account any overhead imposed by the evolution of intelligent life. There may be a minimum time required for evolutionary or geophysical reasons.)

(2) Intelligent life originates through an accident or set of accidents. These accidents are not too unlikely to occur within the habitable lifespan of an Earth-like planet. That means that some planets get "lucky" and evolve observers. With other planets, their sun turns into a red giant before the accidents can occur. When intelligence does arise, it can appear at any point in the planet's habitable span.

(3) Intelligent life is contingent on accidents so improbable that it rarely occurs, even on a suitable planet. When and if observers appear, this is about equally likely to occur at any point in the planet's habitable span.

We now know the age of the Earth to four significant figures: 4.543 billion years. Unfortunately no such precision attaches to early fossils. It's hard to date them and hard to be sure they are fossils. Currently the oldest known fossils are understood to be at least 3.5 billion years old, and conceivably as much as 4 billion years old.

Human-level intelligence, on the other hand, arrived practically yesterday by the cosmic calendar. The anatomically modern human skulls of about 200,000 years ago are in the most recent 0.0044 percent of the planet's history.

We think we know something about the Earth's future. It must be closely tied to the sun's. A star like our sun is expected to shine for about 10 billion years before running out of hydrogen fuel. It will then begin fusing helium into carbon or heavier elements, turning into a red giant star. As a red giant, the sun will engulf the orbits of Mercury, Venus, and possibly Earth. Even if it doesn't quite reach Earth, noon would mean a red sun filling practically the whole sky. Earth would be toast.

That sets an upper limit on how much longer life can survive on Earth. It may seem that we've got about 5 billion years left for Earth's habitable lifespan. But many geophysicists now think this is overly optimistic. The sun is getting gradually hotter even now. In as little as a billion years rising temperatures may evaporate the atmosphere and oceans, transforming Earth into a hot, lifeless planet like Venus. (That's assuming that human-caused climate change doesn't trigger a runaway greenhouse effect much, much sooner.)

As a rough guess, assume the Earth will have about 6 billion years from formation to losing its oceans. The diagram compresses those 6 billion years into a bar. The shaded part shows the fossil history of life on Earth. Life arose early; intelligent life arose fairly late.

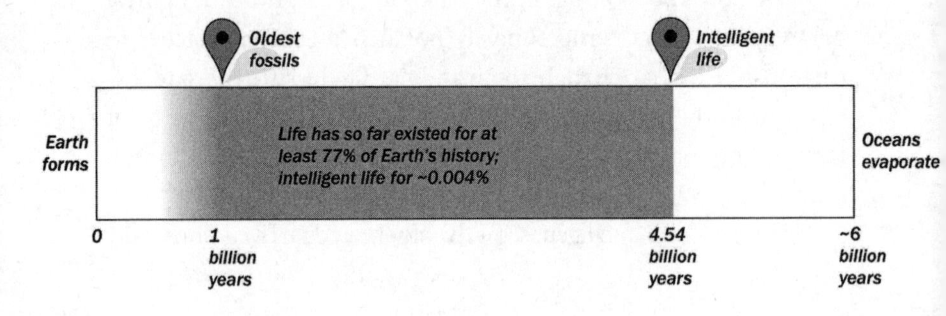

We think that life is just a chemical reaction. Chemists have little use for probabilities. Mix Coke and Mentos. *Whoosh!* They react.

But we don't know the details of how the first self-reproducing units came into being. It's not inconceivable that life was the result of a fantastically improbable accident. It could have been difficult to bring all the needed molecules together in just the right way at the same time in order to get a self-reproducing whole that was not immediately destroyed by something else. This may have been so unlikely that it might not happen, anywhere in a planet's worth of molecules, for billions and billions of years, or not at all. Likewise, it's conceivable that the evolution of intelligence is something extremely improbable.

Can we rule this out, based on anything we know about the history of life? Not really. In terms of the three possibilities I gave above, we can rule out intelligence being a slam dunk (1). Otherwise it's unlikely we would have arisen so late in the Earth's projected habitable span. Our timing is consistent with intelligence being a fairly common accident (2) or a super-rare accident (3).

Carter takes this further. We know of no reason why the average time to evolve intelligence should be related, *at all,* to the average hydrogen-burning lifetime of a sun-like star. The first is a matter of biology, the second of physics. They could be many orders of magnitude apart for all we know. So we can tentatively rule out (2) as a highly unlikely coincidence. That would leave us with (3). In Carter's view we should assume that the evolution of intelligence is highly improbable, and that it is a rare phenomenon in the universe.

Carter's Angel

There was a time when it was imagined that all multicellular organisms, from kale to hookworms to reality show stars, share a common ancestor. We have since learned that this is wrong. The leap from single-cell to multicell life occurred multiple times.

The history of life on Earth includes many other examples of adaptations arising more than once. A striking example is eyes. Anyone who's looked eye to eye with an octopus has experienced a sense of the uncanny. This creature has two eyes with pupils, lenses, retinas, and optic nerves (though no corneas or blind spots). Yet the eyes of the octopus evolved independently of those of fishes and their humanoid descendants. The eyes of insects are unrelated to both. It's believed that eyes evolved about ten distinct times on Earth.

This is provocative because an eye is not just a biological camera. It's part of a biological computer. Creatures with big eyes have big brains, able to map fast-changing data into a continually updated model of the 3-D world. Such creatures are active in evading predators, being predators, and pondering.

In Sagan's 1971 response to Crick, he said that "there are many paths to the origin of life, and...the joint probability that one of them has been taken on a suitable planet over billions of years is rather high." The same might be said of intelligence. The probability of a particular path may be minuscule, but the probability that some path is taken could be high.

But what does that tell us, exactly? It bolsters the case that the evolution of multicellular life, and the evolution of eyes, are not-improbable outcomes *on an Earth-like planet that did, in fact, develop intelligent life.* These adaptations were not roadblocks on the many paths (many random walks) to intelligence.

But there might be other developments, just as crucial, that are highly improbable. As far as we know, life itself arose only once on Earth, and intelligence evolved only once. (All existing life shares the same genetic code. We don't find ruins of intelligent dinosaur civilizations.) Given that life and intelligence each had to arise at least once for us to be here, we are in no position to conclude that they were probable outcomes.

It's hard to be certain what's crucial and what isn't. Suppose, said Carter, that we had evolved wings, like angels. We might now be telling ourselves that wings are indispensable for intelligent life. Otherwise, how would we get around and still be able to use our opposable thumbs?

Carter captured some of these issues in a remix of his fairy tale. Suppose that the suitors must pick not one lock but five within the allotted hour. The five locks must be picked in strict order. Lucky Chuck succeeds, taking a total of 47 minutes and 40 seconds for all five locks.

Chuck may figure he's pretty good at lock picking. He opened all five locks, with time to spare. But once again, he needs to think about the selection effect. Regardless of whether he had to pick one lock or a thousand, the fact that he's still alive mandates that he picked as many locks as he had to—in order to be still alive.

A successful suitor's completion time depends on how many difficult locks have to be picked. Here "difficult" means a lock that, on average, takes much longer to pick than the permitted hour. The more difficult locks there are, the more likely the total time is to be near the end of the allotted hour.

Homo sapiens arrived on the scene at about the 75 percent point in our planet's habitable running time. Carter supplied a mathematical analysis suggesting that the number of critical and improbable steps in our evolution as observers could be as little as one or two. Robin Hanson, an economist at George Washington University, has done computer simulations that put the number at a few more, maybe five crucial steps. Had there been many more, we'd expect to have evolved just in the nick of time, shortly before the Earth became intolerable for life.

The Chaos of Orbits and Climates

Chaos theory says that many phenomena are unpredictable because very small changes in the initial conditions lead to large differences in the outcome. We can never get enough precise data about the current state of the atmosphere to predict the weather very far into the future. Biological evolution is chaotic as well.

Yet we have no trouble predicting eclipses millennia into the future. That's because our solar system runs like clockwork, with all major planets in stable, close-to-circular orbits. We have assumed this is typical. But our fledgling studies of exoplanets, those around other stars, show that our system's harmony is at least moderately unusual. Many of the star systems we've detected elsewhere have planets in highly elliptical orbits, or closely spaced orbits, or orbits not in the same plane. Such orbits would be chaotic, as gravitational pulls and pushes would change them over time.

Computer studies have shown what a complex dance planetary orbits are. Even in our own solar system, the stability may be temporary. Computer models suggest that the Earth and Mars may eventually edge closer and collide, destroying both and creating a new asteroid belt. Planets may be ejected into the frigid outer reaches, or sent spiraling inward toward the sun.

It looks like many or most Earth-like planets eventually get knocked out of their orbits, with temperatures changing so much that any life would be exterminated. How did we manage to avoid this fate? Maybe we just got lucky.

Planetary climates may be chaotic too. The sun is now about 30 percent hotter than it was in its youth. This creates a puzzle, for the Earth is not 30 percent hotter. Liquid water has existed on Earth for more than 4 billion years. It appears that the planet's atmospheric dynamics have changed in order to keep temperatures near their current values, even as sunlight got ever-stronger. It remains unclear how this happened and how likely it was.

The Earth's climate has often been colder than it is now. The Cryogenian ice age, from about 720 to 635 million years ago, produced a "snowball Earth" in which sea ice and glaciers covered the whole globe. This was no brief glitch. It lasted about 85 million years, longer than the Jurassic period. It's believed that volcanoes ended the Cryogenian ice age by pumping carbon dioxide into the air, triggering a global warming that melted the ice and led to the evolution of life as we know it. Lucky for us those volcanoes erupted?

Also lucky that an asteroid or comet slammed into the Yucatan region at the end of the Cretaceous period. It was destructive enough to take the dinosaurs out of the picture, though not so big as to exterminate the vermin that evolved into the writer, and reader, of these lines.

Two Questions for an Extraterrestrial

At Princeton J. Richard Gott III taught a popular introductory astronomy course with Neil deGrasse Tyson and Michael A. Strauss. Gott is a big believer in homemade visual aids. He has found that familiar, tangible objects can often convey ideas better than a digital slide. He showed me one visual aid: an assortment of circular coasters and mouse pads of various sizes. Gott puts Lego-style toy people on the coasters. The coasters are planets, Gott announces. Suppose that one of the little people is *you*. Which planet are you likely to be on?

Everyone can see that they're most likely to be on the biggest planet, the one with the most people on it. And this demonstration, Gott believes, has much to say about why we've never met an extraterrestrial.

Science explains the rainbow we see, not the unicorn we don't. That's the way it normally works, anyway. The Fermi question takes us down the unicorn path. Explain why we *don't* see evidence of spacefaring ETs. Gott's 1993 *Nature* article discussed that issue as well as the survival of our species.

Consider the nations of this Earth (for there we have population figures). You are probably not reading these words in Tuvalu or Liechtenstein or Monaco. The odds are that you are in one of the world's few most populous nations. Together the seven biggest nations by population (China, India, United States, Indonesia, Brazil, Pakistan, and Nigeria) have just over half the world's people.

The median population of the world's sovereign nations is only 8.4 million. Yet practically everyone (96.4 percent) lives in a nation with a greater-than-median population. In that sense, we are almost all "above average."

This is another paradox that isn't. It's a consequence of the fact that nations' populations vary drastically, over many orders of magnitude. China has nearly 10 million times the population of Vatican City, the least populous sovereign state. Consequently, the few biggest nations command most of the world's population. A randomly chosen person is nearly certain to live in a nation that is more populous than most.

Now apply this to the Fermi question. Starting with Enrico Fermi himself, discussions of extraterrestrial life have routinely imagined that colonizing the galaxy is what ETs do. The ETs presumably attain immense populations and exist for millions or billions of years.

Gott says we shouldn't be too sure of that. Start by considering how many observers exist in our galaxy. This number must depend strongly on whether interstellar travel is feasible. If it's *not,* then every observer-species would be confined to its home planet.

Observers would exist only on those rare planets that evolved them. Let's say there are 100,000 intelligent species in our galaxy (a typical Drake equation estimate). A planet might support 10 billion observers. That means the galaxy could have a quadrillion (10^{15}) observers in all.

But if interstellar travel is possible, then we'd expect that some ETs would explore and populate the galaxy, settling on planets that had not evolved intelligent life. The ETs might also develop technologies like "terraforming" that would allow them to render formerly unsuitable planets habitable. This would result in much, much larger populations than would be possible otherwise. There are something like 300 billion stars in our galaxy, and at least a billion potentially inhabitable planets. If we again assume an average population of 10 billion per planet, we find that a fully populated galaxy could have 10 quintillion (10^{19}) observer life-forms.

That's ten thousand times more than the case where species are confined to their home planets. In the scenario where interstellar colonization is possible, practically all observers would be members of a galactic empire.

But we aren't. Here we are, still on our home planet. This makes a case that interstellar colonization has not happened in our galaxy and maybe isn't in the cards. In Gott's view, spacefaring ETs aren't coming here because there aren't many, or any, of them out there.

Just as most residents of Earth live in the few most populous nations, most inhabitants of the galaxy would probably be members of the most populous observer-species. Most likely *Homo sapiens* belongs to that group. Our species is probably above average in cumulative population compared to its galactic peers.

Another of Gott's visual aids explores time rather than space. Gott stacks dominoes to make a histogram of population. Time is the horizontal dimension, and the height of the domino stack represents population (for a given species, such as ours). A random

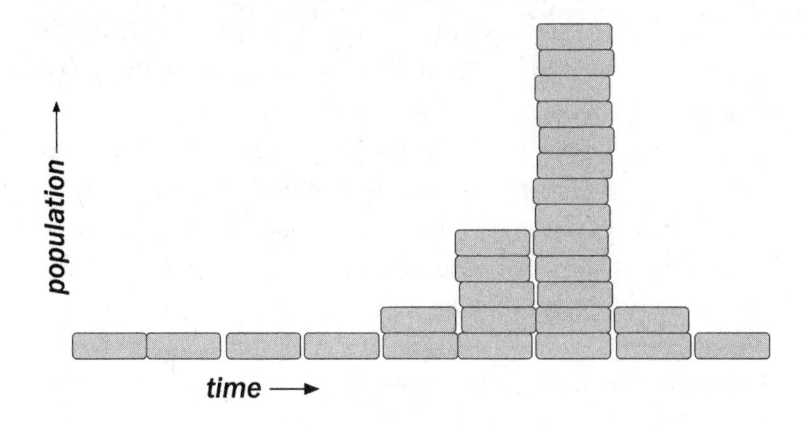

person (domino) is likely to reside in one of the most populous centuries (highest domino stacks), just as she's likely to live in one of the most populous nations.

This illustrates John Leslie's point that it's usual to live during a population explosion. This is true for lemmings, bacteria, or any population that grows exponentially, given unlimited resources. The default assumption should be that we, along with most of the universe's observers, will find our species' population to be presently much greater than it was through most or all of the historic past.

This prepares us to answer Fermi's original question: Why aren't ETs visiting us in their spaceships? Gott says we should start by asking a question we're better prepared to answer: Why aren't *we* visiting ETs in *our* spaceships?

We aren't because we haven't developed the technology. (Yet?) In that we may be more typical than we think.

We should not be too quick to assume, in the absence of evidence, that we're extraordinarily early in the timeline of invention. Gott suggests that it may be typical for an intelligent species to develop technology, experience a population boom, and contemplate many ambitious plans (like interstellar travel) that go unrealized. Right now, in the Gliese 221 system, they could be saying, "We really ought to build starships and explore the galaxy someday. Like in the movies!" But it's all talk.

It's true that many ETs will have had a head start on us of millions or billions of years. They will be that much more "advanced," toward whatever fate has in store for them. But let's not jump to conclusions about what that means. Those old civilizations may have advanced themselves all the way to extinction. We have not one iota of evidence that interstellar travel and multimillion-year civilizations are common things. Gott puts this in Copernican terms.

> If you believe that our intelligent descendants will last 10 billion years and colonize the Galaxy, you must believe that you will, in the end, turn out to have been very lucky to have been in the first tiny fraction of the members of our intelligent lineage.... If you were not lucky enough to find yourself on the first page of the phone book or were not even born on January 1, can you feel comfortable assuming that you will turn out to be even luckier in the ultimate chronological list? You should be suspicious of any claim that future events will conspire to make you *in the end* turn out to be exceptionally lucky, like the claim that you will win the lottery tomorrow or get rich by participating in a chain letter you have received.

We do not have starships. We do have radio. Why don't we hear radio signals from ETs?

On Earth we've been broadcasting a little more than a century. Our radio signals are powered to reach the suburbs, not the stars. The same may go for most ET broadcasts. The Copernican method predicts that we'll use radio technology for another century (median estimate) or 2.6 to 3,900 years (at 95 percent confidence). These figures are shorter than estimates typically plugged into the Drake equation for the lifetime of a communicating civilization. The 1961 estimates were 1,000 to 100 million years.

Could we detect regular broadcast signals from distant stars? It's not likely, given the technology at our end. Our SETI (Search for Extraterrestrial Intelligence) efforts bank more on the hopeful prospect that advanced ETs will want to communicate with us and are able to build superpowerful beacons to do so.

Have *we* ever built such beacons? No. Will we do so? It's hard to say. Already some fret that interstellar broadcasts would paint a target on our back. If our situation is typical, serious attempts at interstellar communication may be rare or nonexistent.

In 1993 Gott reran the Drake equation numbers using Copernican method estimates. He put an upper bound on the number of radio-broadcasting civilizations in our galaxy at about a hundred (at 95 percent confidence). The actual number could be a lot less, even one (us!). Gott's estimate implies that no more than 1 in a billion stars has a transmitting civilization. The nearest such civilization would be no closer than about ten thousand light-years away. That's more distant than any of the stars making up our night sky's constellations.

Is SETI a waste of money? No, says Gott, for two reasons. First, SETI has some nonzero chance of detecting signals. The interest and importance of a positive result may justify the effort, even if the chance is small. Second, the value of a negative result is not to be underestimated.

From 1995 to 2004 Project Phoenix, a SETI effort led by Jill Tarter, targeted 800 stars within a 200-light-year radius of Earth. If Gott's numbers are right, Tarter's chance of picking up ET signals would have been no better than 800 in a billion, or 0.00008 percent. Others surely believed the chance of detecting a signal to be greater than that. Science is about learning the way the world really is. If radio-signaling ETs are rare, that's worth knowing.

* * *

Gott's answer to the Fermi question does not depend on ad hoc assumptions like: interstellar travel is impossible... every ET species annihilates itself in a global holocaust... ETs want to keep us in a zoo... ETs want to exterminate us.... Gott's only assumption is that our vantage point as observers is unlikely to be special.

If our one data point tells us anything, it's that observer-species exist on time scales of hundreds of thousands of Earth years. This stands in contrast to the routine assumption of ET civilization lifespans thousands or millions of times longer than anything in the human scales of history. It is this unsupported and often unexamined assumption that is a root of the Fermi paradox, Gott believes. Better to look the modest evidence we've got straight in the eye.

Our ongoing and valuable SETI efforts may prove to be an "emperor has no clothes" moment. Perhaps most ETs do not, after all, go on to explore many planets, achieve immense populations, and proclaim their existence to the whole universe. We've got a nice planet here. For most of the universe's species, that may be all there is.

Gott says that spacefaring ETs are not nearly so common as is generally assumed. If he's right, then it seems that typical Drake equation estimates are wrong. How can they be so far off? A 2018 paper by Anders Sandberg, Eric Drexler, and Toby Ord, all of the Future of Humanity Institute, Oxford, provides a convincing answer. In the Drake equation we are not just multiplying seven unknowns; we are also multiplying the uncertainties in those unknowns. Any Drake estimate inherits all these uncertainties. But we tend to overlook that. We plug in values and get an answer like 10 million ET civilizations in our galaxy. We tell ourselves it's a just a guesstimate.... And having said that, to appease the gods of statistics, we thereafter wonder why reality doesn't seem to agree with our estimate. But given the uncertainties that underlie this figure, reality can be very different from the estimate, and probably is.

Here's a simplified example of what the Oxford group is talking about. Suppose there are three kinds of galaxies in the universe, all equally common. One kind has 1 million ET civilizations, another kind has just one, and the third kind has 1/1 million (meaning that the chance of even one is just 1 in a million, so there probably aren't any). We don't know what kind of galaxy ours is. How many ET civilizations is it likely to have?

Well, there's a one in three chance that our galaxy is the kind that harbors a million civilizations. That averages out to 333,333. The other two galaxy types don't add much to that, so 333,333 is the mean or expected value for the number of civilizations.

But this "average" is much greater than the median outcome, the one in the middle. The median case is that our galaxy has just one civilization. That's a big difference. And more to the point: the fact that our galaxy has 333,333 civilizations "on average" doesn't exclude a very real possibility (a one in three chance) of it having no ETs at all.

Actual Drake equation estimates are even more skewed than that example. Sandberg and colleagues collected estimates that had appeared in the scientific literature. They then did a computer simulation in which the value for each factor was chosen by drawing randomly from among the published estimates for that factor. Seven chosen factors were multiplied together to get a virtual Drake estimate, and this process was repeated many times to reveal the range of variation. The Oxford group found that the resulting estimates varied by more than forty orders of magnitude. This is mainly due to uncertainties about the probability of life arising. Many now side with Brandon Carter, that life could be an incredibly rare accident. Uncertainties about the lifetime of a communicating species are second to that.

In the Oxford simulation, the mean number of ET civilizations in the galaxy was a generous 53 million, though the median was only 100. But, because of the spread-out distribution, there was nonetheless a 30 percent chance of *zero* ETs in the galaxy. Indeed, about 10 percent of the estimates implied that ET civilizations are so rare that there are unlikely to be any in the observable universe.

Accept this, and there is no Fermi paradox. We have no reason to be surprised at the absence of ETs. We may well be alone in the galaxy or even the universe, and it's not out of line with what scholars currently believe. Where is everybody? The Oxford group's answer is "probably extremely far away, and quite possibly beyond the cosmological horizon and forever unreachable."

Lemmings and Black Swans

There are optimists who say that the past is no guide to the future. They speak of a singularity. This can be raised as an objection to Gott's analysis. If most ET civilizations are not too different from our own, then they will experience rapid technological advancement. Assume only that some ETs are a few millennia ahead of us (which is *nothing* in cosmic time), and they already could have experienced a transformation in which technologic power zoomed abruptly upward. *Anything* could be possible (like faster-than-light travel?). So even if most ETs are slackers like us, and even if the nearest civilization is billions of light-years away, we shouldn't underestimate the power of sufficiently-advanced-technology-that-is-indistinguishable-from-magic (in Arthur C. Clarke's words).

But look at what the Copernican philosophy is actually saying. It is reasonable to think: *I observe myself to exist during a time of rapid technological advancement; therefore, I infer that many observers will share that observation. Typical ETs may also find themselves in a time of technological advancement.*

Yet the fact of my living in a boom time does not, by itself, tell me how long this technological advancement will continue, or what heights it will ultimately reach, on Earth or anywhere else.

The lemming analogy may be instructive. A thoughtful lemming would observe that the flock's discoveries of berries, moss, and lichen have been growing exponentially. It would be right in thinking that most sentient lemmings, throughout the Arctic, experience the same thing. It would also be justified in predicting that there will be exponentially more berries and moss discovered next week, *provided there is a next week for the flock.* It can go on reasoning this way right up until the flock leaps off that mythical cliff.

So yes, it *does* follow that some ETs will be much more advanced than we are. But these super-advanced ETs may be much rarer than we think. What exponentially advanced technology means for population and survival is an open question. Today many worry that our accelerating digital technology will be the agent of our destruction. As Gott wryly puts it, Einstein was a smart guy, but he didn't live any longer than a lot of people who weren't so smart.

In Ulam and von Neumann's original sense, the singularity is the ultimate black swan event, not to be predicted from anything that came before, and "beyond which human affairs . . . could not continue." The singularity could be another name for doomsday.

Gott told me that, if he ever met an extraterrestrial, he would immediately want to ask two things. One, *How long has your civilization lasted?* The other, *How similar are you to us?* For instance, Gott would want to know whether the ET's language had a word for "war." The ET would offer that elusive second data point, a second random draw from the urn of possibility. It would go a long way toward revealing whether the human condition is universal or unique.

Pandora's Box

Buried in the mountainside near Geneva, CERN's Large Hadron Collider (LHC) is the most powerful particle accelerator ever built. It is a circular vacuum tunnel in which magnets accelerate protons to nearly the speed of light. The protons smash into one another head-on, producing bursts of other particles. High among the LHC's goals was to discover whether the Higgs boson exists. But the collider's inauguration in late 2008–2009 was marred by a series of unfortunate events.

One was unwelcome publicity. A handful of bloggers talked up speculation that the LHC's unprecedentedly powerful collisions might create micro–black holes. Smaller than an atom, these hypothetical points of mass would fall through solid rock like it was air. Subject to gravity, the micro–black holes would revolve around the Earth's center in elliptical orbits. With each hungry pass they would sweep in more of the planet's mass. Ultimately, it was said, the whole Earth might be consumed by an ever-growing black hole.

Though the hazard was dismissed by nearly all physicists, the consensus talking points shifted even as the LHC was being built. It had been denied that micro–black holes were even possible....

Then certain results in string theory suggested they might be. It was asserted that any micro–black hole would instantly evaporate through Hawking radiation. Further work threw that into question. It was said that planet-eating micro–black holes, if possible, would also be created in white dwarf stars. We see no evidence of them there. This string of conflicting ideas is typical of scientific inquiry. But with the apparent stakes so high, it inspired gallows humor. There were cleverly apocalyptic T-shirts for purchase and worried souls checking the news the day the LHC was switched on.

On September 10, 2008, the collider began preliminary tests at relatively low energy. Nine days later, in the so-called quench incident, six tons of liquid helium coolant escaped into the vacuum pipe with explosive force, damaging sensitive equipment and halting operation. Repairing the damage delayed the LHC's scientific program by more than a year.

In October 2009 a CERN physicist was arrested for ties to Al-Qaeda. The following month, a piece of a baguette, apparently dropped by a bird, short-circuited outdoor equipment. It might have led to a second quench incident had the LHC not still been off-line from the first.

During the downtime, physicist Holger Bech Nielsen, one of the original doomsayers, and colleague Masao Ninomiya began posting yet more bizarre speculations about the LHC's troubles to an internet site. These too were quickly picked up by the media, resulting in such headlines as GOD IS SABOTAGING THE LARGE HADRON COLLIDER and TIME-TRAVELING HIGGS SABOTAGES THE LHC. NO, REALLY.

In effect the physicists were proposing an observation selection effect. Suppose that operation of the LHC would cause a catastrophe ending all earthly life. Then, given that we're alive, we necessarily live in a quantum world in which all attempts to create fatally powerful colliders have been prevented by a series of "accidents." The quench incident, the Al-Qaeda physicist, and the piece of French bread were part of a cosmic conspiracy preventing the collider's operation.

Nielsen and Ninomiya ("a pair of otherwise distinguished physicists" in one journalist's assessment) even dragged the US Congress into it. In the late 1980s both the United States and the Soviet Union planned and started building their own supercolliders. It was an underground "space race." With the fall of the Soviet Union, work on the Russian collider was halted. In turn, budget-cutting Republicans in the US Congress decided the United States no longer needed an expensive scientific initiative. The American collider plans were scrapped, not long after the Berlin Wall was.

"Our theory suggests that any machine trying to make the Higgs shall have bad luck," Nielsen said. "It is based on mathematics, but you could explain it by saying that God rather hates Higgs particles and attempts to avoid them."

The LHC saga posed two questions. First, is it possible that a physics experiment might inadvertently destroy the whole world? Second, is it possible that quantum physics would somehow prevent such a catastrophe? Both have been explored from the standpoint of Bayes's theorem, self-locating evidence, and selection effects. I'll treat the first question in this chapter, and the second in the next.

Taking the Fermi paradox at face value, something has prevented or curtailed the emergence of space-traveling and communicating species. This hypothetical something is often called the *great filter.* "It is not far-fetched to suppose," wrote Nick Bostrom, "that there might be some possible technology which is such that (a) virtually all sufficiently advanced civilizations eventually discover it and (b) its discovery leads almost universally to existential disaster."

That's one possibility. But the great filter need not be one thing. The term can be used to embrace everything affecting the Drake equation factors. We can view the past only through the rose-colored lenses of our selection effect. Here we are, and we don't know how much luck it took to get here.

That is a sobering thought. It might be that earthly life has been on a long lucky streak up to now, and that streak is about to end.

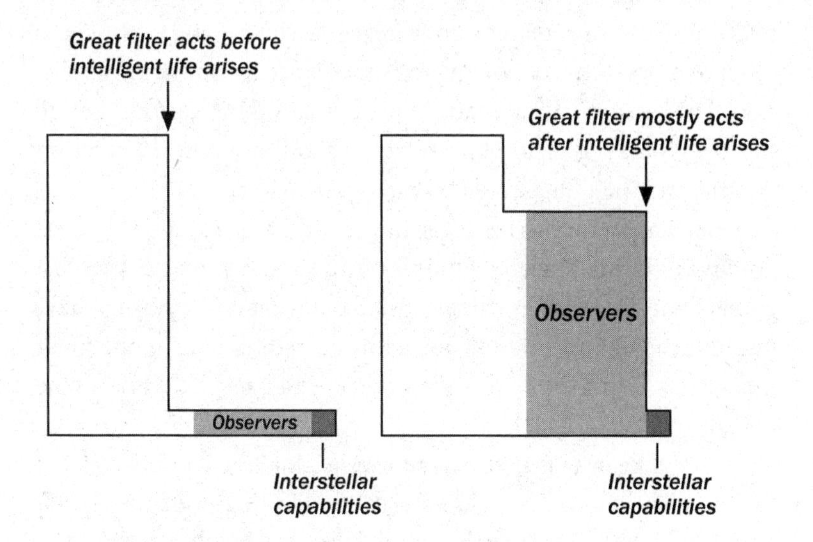

The diagram shows two possibilities. Each is a domino-stack histogram of planets suitable for life. Time moves from left to right. In both scenarios, we start with many habitable planets (high stacks at left) and winnow that down to just a few with observers (shaded region). A few of those observers attain the ability to communicate and travel over interstellar distances (dark shading). The regions of dark shading are the same size in each diagram. This is the Fermi constraint: communicating civilizations must be rare, or we'd probably already have detected some.

Shown are two ways of arriving at a world with few communicating ETs. In the scenario at left, the great filter mostly acts before observers arise. This might be because a molecular miracle is needed to jump-start life, or because the long-term instability of planetary orbits almost always prevents the evolution of intelligence. We would then live in a universe almost empty of observers. We could count ourselves incredibly lucky to be where we are. We would also be lucky in that the filter is behind us. We have a clear path ahead to attaining interstellar capabilities.

The diagram on the right shows a more forbidding case. It is relatively easy for planets to evolve life and observers. These observers develop technology and experience a population explosion. The filter mostly acts after the population increase but before the technology gets to interstellar capabilities. The population booms, and then the population goes BOOM?

Say I'm a random observer in the universe, occupying a random point in the shaded region(s). With the left scenario, there's a good chance my species will eventually make it to the stage of having interstellar capabilities. But with the right scenario, the great filter lies ahead of me, and the odds are my species won't make the cut.

One takeaway is that maybe we shouldn't be rooting so hard for the discovery of extraterrestrial life. Finding even simple life anywhere other than Earth would increase the chance that the great filter is something in our future, not the past. "It would be great news to find that Mars is a completely sterile planet," Bostrom said. "Dead rocks and lifeless sands would lift my spirits." That's because finding life on Mars, Europa, or Enceladus would greatly boost the odds that life is abundant in the universe. It would be evidence that the early stages of life's development are easy—it's the last step that's a doozy.

Salmonella

John Leslie is a connoisseur of catastrophe. He collects news clippings bearing on novel and overlooked ways the world might end. It is not a small collection. A while back Leslie came across reports of a new contraceptive. Salmonella bacteria had been genetically engineered to produce a benign infection that rendered women infertile for a few months. This was reported as another miracle of modern medicine. But salmonella is everywhere, the hobgoblin of unwashed cutting boards and potato salad left too long in a summer trunk. What, Leslie wondered, if the contraceptive strain escaped and mutated, to become extremely contagious, with permanent infertility? The Omega strain is otherwise harmless. No one gets sick, and no one dies. But no one thereafter has babies, ever. A century later, the human race is extinct.

That's just a for-instance. Leslie has plenty more. The bigger point is that, for a species able to tinker with genetic material, its first big mistake could be its last. Is it realistic to think we can develop genetic engineering without making a few mistakes along the way? Some of them really *big* mistakes?

Any answer to that question must consider biohackers, a subculture of hobbyists not bound by the safety and ethical guidelines of academic biology. Biohacker Josiah Zayner claims he was the first person to try to modify his own genome with CRISPR, the gene-editing technology. In a 2017 article Zayner was quoted as saying, "I want to live in a world where people get drunk and instead of giving themselves tattoos, they're like, 'I'm drunk, I'm going to CRISPR myself.' "

The Metastable Vacuum

Other disturbing thoughts center on unfortunate physics experiments. The protons and neutrons of atomic nuclei are themselves composed of smaller particles known as quarks. Quarks come in six varieties ("flavors"), given the arbitrary names *up, down, strange, charm, top,* and *bottom*. The important thing is that all familiar matter contains up and down quarks only. The other flavors of quarks are known only from collider experiments and always vanish a split second after their creation.

Unless... There is a hypothetical substance called *strange matter.* It's theorized that strange quarks may be able to combine with the up and down quarks of atomic nuclei to form stable particles ("strangelets") and an exotic form of matter. In the worst-case scenario this strange matter might be like Kurt Vonnegut's "ice-nine." The merest speck of it coming into contact with ordinary matter would convert the whole Earth into strange matter. This Midas-like transformation would destroy anything made of atoms, such as us. Perhaps a powerful collider will one day produce a strangelet.

Strange matter is by no means the worst possibility. Another conjectural collider by-product is a *vacuum metastability event.* A vacuum is said to be a space containing nothing at all. Yet what we can so easily imagine doesn't exist. Quantum theory says that space cannot be completely empty, even in principle. In the emptiest possible space, virtual particles briefly flicker into and out of existence. These particles propagate fields such as those of electromagnetism or gravity. A vacuum must therefore contain a small residue of energy.

It is generally believed that our familiar vacuum contains the smallest amount of energy possible. But in 1980 physicists Sidney Coleman and Frank De Luccia proposed that another, lower-energy kind of quantum vacuum might be possible. If so, our familiar vacuum would be only "metastable," liable to transform at any moment into the other kind of vacuum. Should even a tiny bubble of that lower-energy vacuum come into being, say in a collider experiment, it would be "the ultimate ecological catastrophe," the authors wrote. Our familiar vacuum state would instantly convert to the lower-energy state, releasing massive amounts of energy and destroying everyone and everything we care about.

A strange-matter catastrophe would perhaps affect only the Earth. A metastable vacuum event would affect all of space and future time. It would propagate outward in all directions at virtually the speed of light. The ever-growing vacuum bubble would consume planets, stars, galaxies, everything. It wouldn't just be the end of us. It would be the end of any ETs out there. Inside the bubble there would be new constants of nature incompatible with life. Protons would instantly disintegrate. Matter would quickly undergo gravitational collapse. In words unusual for the *Physical Review D,* Coleman and De Luccia wrote that it would be the end not just of life as we know it but of any "structures capable of knowing joy."

It must be emphasized that we don't know whether a lower-energy form of vacuum exists. It's pure speculation. (As were black holes, X-rays, and electricity.)

In 1983 Piet Hut and Martin Rees published a paper in *Nature* assessing the metastable vacuum risk posed by collider experiments. It is believed that only very energetic collisions could create a lower-energy vacuum. Hut and Rees noted that natural cosmic rays produce particles far more energetic than those in colliders. Cosmic ray energies go up to about 10^{11} GeV (giga-electron volts). Collider energies were only about 10^{3} GeV at the time (and 10^{4} GeV now). Thus our colliders can hardly increase the overall risk, given that the universe is awash in cosmic rays.

Maybe this is true, for now. But it kicks the can down the road. Someday we may have the ability to create collision energies never realized in nature. Do we then stop building particle accelerators?

It is easier to speculate on human nature than on physics. There will be a group of physicists who say the next-generation experiment is safe, and a group who say it might not be. How the dispute will play out will depend on the relative size and prestige of the factions, plus how the cases are presented in the media. Ultimately, neither the public nor political leaders will have a deep understanding of what the physicists are saying.

So maybe the experiment is outlawed as being too risky, just to be on the safe side. We've only got one planet, one universe. Years later the Mars colony wants to demonstrate how innovation-friendly it is. It approves the hyper-super-accelerator, to be built on Martian soil. Yet if the experiment undermines a metastable vacuum, it will be just as lethal to Earth as to Mars. It is easy to see the collider experiment as a Pandora's box. Sooner or later, somebody's got to open it.

Note that we have to worry not just about *our* physics experiments but about those of ETs. Yesterday, in a galaxy far, far away, someone might have switched on a brand-new 10^{20} GeV particle collider. The end. It could have instantly created a bubble of lower-energy vacuum, destroying the collider, the planet, and its sun. The bubble would now be racing outward, one day to reach the Earth. Since it's traveling at the speed of light we would have no advance warning. We would witness the faraway galaxy's destruction at the precise moment of our own.

This suggests an ironic answer to the Fermi question (which shares its namesake with the Fermi National Accelerator Laboratory). All the ways that we might learn of ETs—radio beacons powerful enough for interstellar communication, spaceships capable of reaching another sun, and engineering on a cosmic scale—involve incredible amounts of energy. A civilization able to do these things would also be able to build superpowerful particle accelerators. It may be that ETs almost always do a fatal experiment before they get around to exploring the galaxy.

And if that experiment produces a metastable vacuum event, the universe might be riddled like Swiss cheese with ever-expanding bubbles of vacuum. We would inhabit the dwindling interstices between vacuum bubbles, an ever-shrinking world in which everything seems to be okay. When the end comes, we won't even have a split second to gape at the sky and say "WTF?" There will be a last moment of consciousness, then nothing.

* * *

"There is something so clean about such complete annihilation, analogous to vanishing in a genie's spell," wrote William Eckhardt of this idea. He observed "that people do not feel the emotional impact that they do from contemplating slower deaths by nuclear holocaust or ecological disaster."

Bostrom makes a similar point. We've never witnessed anything like this (obviously), so we don't have any cognitive infrastructure for dealing with it. Meanwhile, there are so many other clear and present dangers demanding our attention.

Bostrom also warns of what he calls "good-story bias." We favor notions of the future that make a good story. That means that catastrophes should be the result of tragic flaws, properly foreshadowed, and should not come as non sequiturs. It is easy to accept conforming narratives of the world to come, and to resist alternatives in which the human plotline jumps the shark.

Deadly Probes

Von Neumann probes, those procreating gizmos sent out to explore the galaxy, are usually conceived as benign. Either they go about their business without disturbing anyone, or else they offer greetings to any life-forms they encounter. But in the deadly probes scenario, outlined by Glen David Brin, the universe has a few bad apples. A few observer-species are xenophobic, paranoid, and malevolent. They do not build nice robots. They build von Neumann terminators, designed to exterminate every other species they encounter.

A self-reproducing machine could be weaponized to reproduce without limit, crowding out all native life-forms or turning the planet into nanotechnological gray goo. Or, more efficiently, the probe could sequence the local genomes and devise an artificial virus to exterminate everything.

Why would intelligent beings be so malicious? They might make the Copernican assumption that all ETs are like them, in which case it's kill or be killed. They might be expansionists who view the galaxy as their manifest destiny. When you consider what humans have done to fellow humans to acquire a few square miles of territory, it is not so hard to imagine the remote-control extermination of aliens who are "not like us."

We don't have to assume that bad ETs are common. It only takes one. Once the deadly probes are unleashed, there is no calling them back (even if, with poetic justice, the originating species meets an early doomsday).

In their explorations, deadly probes might prioritize star systems with intelligent life. They could listen for the radio transmissions of newly technological civilizations and launch themselves toward them. As Brin observes, "*I Love Lucy* has spread well past Tau Ceti by now."

A Timetable of Universal Calamity

It will now be clear that there is no shortage of people able to imagine exotic calamities. The Fermi paradox hints that we shouldn't be too sure such things can't happen. Is there any rational way to evaluate these possibilities?

Max Tegmark and Nick Bostrom address that question in a 2005 article in *Nature*. They begin by acknowledging the selection effect "that precludes any observer from observing anything other than that their own species has survived up to the point where they make the observation. Even if the frequency of cosmic catastrophes were very high, we should still expect to find ourselves on a planet that had not yet been destroyed." This, they warn, is apt to give "a false sense of security."

We can avoid that problem by looking at evidence from celestial objects whose destruction wouldn't have precluded our own existence. Say we're concerned that natural, high-energy cosmic rays might occasionally create micro–black holes that consume whole planets. It could be that this happens to most life-bearing planets (or to their suns), and the Earth has just been lucky so far. Yet our own solar system provides considerable evidence that this isn't a probable or common event. Had a micro–black hole been created in Neptune's atmosphere, it might have long since converted the whole planet to a black hole. A Neptune-mass black hole would be about six inches across, smaller than a honeydew melon. But what happens on Neptune doesn't adversely affect Earth or its life. Black hole–Neptune would still orbit the sun at the same distance and have a gravitational field. Urbain Le Verrier and John Couch Adams would have noted irregularities in the orbit of Uranus and predicted an eighth planet. Johann Gottfried Galle would not have been able to see black hole–Neptune through his telescope. But eventually someone would have discovered something more incredible: a set of satellites orbiting an invisible "planet." We knew of black holes by the time of the Voyager 2 mission (1989) and would have been very interested in the one in our backyard.

But in fact we know that neither Neptune nor any of the other major bodies in our solar system is a black hole. We also know of exoplanets that eclipse the light of their distant suns. Were these planets black holes they'd be too small to eclipse their suns' light. This confirms that planets do not commonly turn into black holes by any natural process.

We can similarly rule out strange-matter planets as a common thing. Spectra of planetary atmospheres show familiar chemistry. The moon landers and the Mars rovers did not transmute into a blob of strangelets. Every working instrument we've put on another planet proves those bodies are made of atoms with nuclei of protons and neutrons—made of up and down quarks.

Another concern is that supernovae might periodically sterilize all the star systems around them. Again, the Earth might have been lucky to avoid this fate. Here Mars and Neptune must share our luck. We can't learn anything about supernova bombardment from within our solar system.

Fortunately we don't need to. We are able to observe supernovae throughout our galaxy and in other galaxies. This gives reasonably accurate, ever-improving estimates of how often supernovae occur and how powerful they are. These estimates say they probably aren't a major obstacle to the evolution of intelligent life. There too we've got data independent of our own existence.

The metastable vacuum and deadly probes scenarios are more difficult to evaluate. It's no good saying that the Andromeda galaxy seems to be free of vacuum bubbles. We would have to observe that—up until the moment the deadly vacuum reaches us.

In principle we could learn about deadly probes via a successful SETI effort. If we were able to listen to one or more ET broadcasts, and none of them mention deadly probes, that would be Bayesian grounds to discount the idea. And if one of those broadcasts turns into *The War of the Worlds*—"Yikes, the deadly probes have arrived!"—we would at least have advance warning. (That's assuming deadly probes travel at sub–light speed. Alternatively, maybe Fermi was right, and all advanced ETs invent faster-than-light travel. In that case the deadly probes could be here, like, yesterday.)

Tegmark and Bostrom realized that we have one data point of self-locating evidence. It is our position in time.

They focus on the formation of habitable planets, those suitable for life and observers. This is something that is reasonably well understood by astronomers (probably more than the evolution of intelligent life is understood by biologists). The first generation of stars did not have the heavy elements required to form rocky planets. Those early stars created heavy elements such as iron and carbon that were recycled into later solar systems. The

Earth formed about 9 billion years after the big bang. This is believed to be a typical time for a habitable planet to come into existence, neither remarkably early nor remarkably late.

That's in the standard astrophysical understanding, which of course does not take into account the kind of weird ideas we're considering here. But let's play devil's advocate. Assume that our vacuum is metastable, with bubbles of lower-energy vacuum arising randomly through space by some sort of natural cause. If the bubbles are common, consuming nearly all of space in a few billion years (say), then observers would be very rare. Practically all the observers that managed to evolve would have to do so early in the universe's history, while there was still a lot of undisturbed space left.

We find ourselves 13.8 billion years after the big bang. That is not especially early (even allowing for the time required for heavy-element formation and biological evolution). Our position in time allows us to put some Bayesian limits on how common metastable vacuum events can be (assuming they exist at all).

Tegmark and Bostrom estimate that the typical time scale for randomly arising cosmic catastrophes must be greater than 2.5 billion years, at 95 percent confidence. That means that, even if vacuum bubbles exist, they are unlikely to reach us for billions of years.

More recently Nima Arkani-Hamed and colleagues used the Standard Model of particle physics to address the issue. They had good news and bad news. The bad news is that they believe our vacuum *is* metastable. The good news is that vacuum bubbles probably won't hit us for about 10^{138} years, a duration so long you might as well call it "forever." That incredibly high value is of course consistent with the Bayesian math (which only sets a lower limit).

Tegmark and Bostrom's reprieve applies to external catastrophes, not those we may create for ourselves. It further assumes those external catastrophes are distributed randomly through space and time. This would not be the case with catastrophes unleashed by ETs. Suppose the metastable vacuum scenario never occurs naturally (as Arkani-Hamed's group is essentially saying) but can occur as the result of ET physics experiments that go awry. Then the risk would be proportional to the distribution of technological species over time. There would have been few ETs, and little risk, in the early universe. It would then be somewhat less remarkable to find ourselves living so late as we do. We cannot rule out ET-created catastrophes with the same degree of confidence.

That applies to deadly probes as well. And if they zero in on our radio broadcasts, then they are correlated not only with the existence of evil ETs but with our own existence. Deadly probes is one weird idea we can't exclude, at least not this way.

But Tegmark and Bostrom's overall conclusion is reassuring: "Our basic result that the exogenous extinction rate is tiny on human and even geological timescales appears rather robust."

Life and Death in Many Worlds

Surely God does not play dice with the universe," said Einstein. Less well known is Niels Bohr's comeback zinger: "I think Einstein should stop telling God what to do."

Both were speaking of quantum theory. In Bohr's view, quantum theory tells us our world is ruled by chance. Einstein found that unacceptable. His famous objection was echoed uncannily in 2009, at the inauguration of the Large Hadron Collider. Holger Bech Nielsen and Masao Ninomiya offered a modest proposal to save the world. We would play cards with God. Or the God particle.

Conceptually, Nielsen and Ninomiya were suggesting something like this: Make up a deck with a million cards. All but one of the cards have blank faces. The other is the Joker, and it says, SHUT DOWN THE LARGE HADRON COLLIDER IMMEDIATELY. We shuffle the deck and draw one card. As long as the card is blank, the European science agency CERN continues with efforts to bring the LHC online and probe the secrets of the universe. But if the Joker is drawn, then CERN shuts down the LHC permanently. For the game to work—to save the world, that is—CERN's leadership would have to agree to abide by the drawn card. It couldn't be a bluff.

The LHC cost $4.75 billion to build, paid by a politically complex consortium of European nations. Not surprisingly, CERN passed on the card game. The last thing they needed was any more doomsday media. On July 4, 2012, the LHC announced it had found a Higgs boson, which the media called the "God particle."

The LHC card experiment is one of the more colorful examples of an attempt to understand the nature of quantum reality. At issue is the "many worlds interpretation," identified in the popular imagination with parallel universes. Many worlds has, however, become a serious scientific question, and self-sampling plays a role in ideas for testing it.

Collapse of the Wave Function

Erwin Schrödinger warned a 1952 lecture audience that they were about to hear something that would "seem lunatic." He announced that we inhabit one of an uncountable number of quantum worlds. This statement was rooted in the work that won Schrödinger the Nobel Prize, work that is the foundation of quantum physics. In barely half a century, Schrödinger's opinion about parallel worlds went from being one man's harmless delusion (as many charitably saw it) to one of the most dogged controversies of contemporary physics.

There is much we seem to be unable to know, measure, or predict about the subatomic world. Werner Heisenberg's uncertainty principle holds, for instance, that we may measure the position or velocity of an electron but not both at the same time with unlimited accuracy. Quantum theory forces us to settle for probabilities, not certainties.

Schrödinger devised the *wave function,* quantum theory's description of the world. For any given situation, the wave function describes the probabilities of its possible outcomes. The wave function decrees, for instance, how likely a Geiger counter is to detect a gamma ray photon (and "click") at a given moment. Once a particle is observed, the wave function is said to "collapse." Like a bursting bubble, the wave is no more. These are figures of speech; the observable reality is that the measured particle is found to exist in one specific state. That could be a specific position or velocity or spin.

Physicists, like everyone else, have struggled to make sense of this. Schrödinger struggled. You've heard of Schrödinger's cat, the cat meme of quantum weirdness. A cat is sealed in a box with a flask of poison. A device, triggered by radioactive decay, has a 50 percent chance of smashing the flask, releasing cyanide gas that will quickly kill the cat. Is the cat, which no one can observe, alive or dead?

In many popular tellings of the tale, the cat is both dead and alive—until the box is opened. Schrödinger did not believe that. He invented the cat experiment as a reductio ad absurdum, a demonstration that our understanding of quantum physics must somehow be wrong.

Another conventional gloss on quantum theory makes a most un-Copernican claim: The observer is special. It is the conscious mind's act of observation that makes the wave function collapse—that makes the poor cat either dead or alive but not both. This credo links mind and matter in a way that has long appealed to mystics but makes most physicists profoundly uncomfortable.

Everett's Many Worlds

Beginning in the 1950s Schrödinger's "lunatic" idea of parallel worlds was embraced and extended by Hugh Everett III and Bryce Seligman DeWitt. It is now known as the many worlds interpretation (MWI) of quantum theory. Many worlds is not easily put into plain English without some distortion, but what it says, roughly, is that all the wave function's outcomes are real. The act of making a quantum measurement is a branch point. There is a branch or world in which a proton's spin is observed to be in one direction, and another branch in which it is in the opposite direction. The wave function never collapses, and it evolves through time with classical determinism. There are parallel worlds with alternate versions of you and me.

Over the past half century many worlds has gained support to the extent that informal polls show a sizable fraction of working physicists saying they believe it. This is not because they think parallel universes is a cool idea (though some do). It's mainly because we now have a better understanding of quantum *decoherence*. Particles display the notoriously bizarre quantum behavior (such as existing in two states at once) only while they are completely isolated from their environment. Once a particle interacts with the randomness of the world around it, it loses its quantum superpowers ("decoheres"). It is found to exist in one place at one time, like a billiard ball or a planet.

Decoherence debunks the mystic role of the observer. It's not the observer but the environment that causes the wave function to collapse. That's because humans are too big and slow to interact with subatomic particles directly. We need an intermediary, such as a Geiger counter scaled to human hands, eyes, and ears. When a gamma-ray photon enters a Geiger counter detector, it ionizes the gas, creating a small current that is amplified by electronics into an audible click. It is the photon's interactions with random gas atoms that constrain it to exist at a particular point in space and time. The human observer has nothing to do with it. He could be texting or taking a nap. It is only narcissism that makes us think the wave function's collapse is all about us.

Quantum decoherence explains why Schrödinger's cat is only a metaphor. A cat is also too big and slow to ever be isolated from the jittery white noise of quantum reality. Only when we consider subatomic particles and nanoscales of time does it makes sense to speak of isolated systems. No man, or cat, is a quantum island.

* * *

In science fiction, characters jump into a parallel universe, explore it, and return to this one. This is not something that's possible with Everett's many worlds. Those worlds are said to be unobservable and unfalsifiable. One skeptical catchphrase is "shut up and calculate." Many worlds, say critics, gives the same predictions as regular quantum theory. It is therefore not a proper theory but only an interpretation, a different way of *talking* about what lies behind the math. The insinuation is that the math is all a serious physicist ought to care about.

On this point attitudes are changing. There are now so many provocative "interpretations" of quantum theory that physicists would welcome an opportunity to toss a few in the recycle bin. Given how many physicists now endorse many worlds, an experiment or observation proving it wrong would be an automatic Nobel. All the more so for an experiment proving it right.

This goal, as physicist Andrew White remarked, has been "like a giant smooth mountain with no footholds, no way to attack it." Despite that, speculative schemes for experimentally distinguishing between many worlds and single-history conceptions of quantum theory are published with some frequency. There is often disagreement about whether the schemes would work, even in principle. Nearly all involve technology well beyond what's presently available. One exception is a low-tech device, the quantum suicide machine.

Quantum Suicide Machine

Max Tegmark has led a charmed life. In high school in Sweden he made a small fortune coding a video game. He's now an MIT cosmologist whose work has been funded by Elon Musk. Yet a share of Tegmark's popular fame is due to some half-joking remarks he made in 1997.

Like a lot of other physicists, Tegmark had racked his brain trying to come up with an experimental test of many worlds. He eventually came up with a notion known as a quantum suicide machine. In David Papineau's words, the high concept is: *Get in the box with Schrödinger's cat.*

Well, there doesn't have to be a box. Tegmark imagines his suicide machine as an automatically firing gun controlled by a device that measures the spin of a quantum particle once a second. Should the spin be "up," the gun fires. Should the spin be "down," the gun merely makes a *click*.

By design the chance of firing is 50 percent at each measurement. The gun therefore produces a random staccato of clicks and bangs. On this all agree.

What happens if you point this same gun at your temple? Like most things in our world, this is a matter of perspective. Anyone who is not you will observe the same random staccato. The first bang will be muffled by the bullet firing into your brain.

Your chances are grim (say the kibitzers). They expect you to have a 1/2 chance of surviving the gun's first quantum measurement, a 1/4 chance of surviving the second... 1/8, 1/16, 1/32, and so on. The chance of surviving a full minute is less than 1 in a billion billion.

Under the single-history version of quantum theory, endorsed by Bohr and deplored by Einstein, there is only this one world, ruled by ruthless chance. It's the surest of bets that anyone who points the gun at his head is going to be killed within seconds.

But according to Everett's many worlds interpretation, there are world branches for each tick of the quantum clock. At each measurement, there's a branch in which the volunteer survives, and a branch in which he doesn't. No one can experience being dead. Any experiencing that's to be done will happen in the quantum worlds in which the volunteer survives. That means the volunteer hears *click... click... click...* and never a bang. No matter how many world branches there are, there is always a version of the volunteer's consciousness that has survived the suicide machine.

Should the volunteer point the gun away from his head, it would resume the random sequence of clicks and bangs. But as soon as he points it back at his head, it's *click*...*click*...*click*...And the longer the volunteer survives, the more confident he can be that the many worlds interpretation is correct.

This is a thought experiment, to think about and not to do. In 1997 Marcus Chown, writing for *New Scientist*, quizzed Tegmark about this. Would he really do such an experiment himself? "I'd be OK," Tegmark replied, "but my wife, Angelica, would become a widow. Perhaps I'll do the experiment one day—when I'm old and crazy."

Tegmark found he was not the first to describe a quantum suicide machine. Similar ideas had been independently conceived and published by Hans Moravec and Bruno Marchal. Moravec, a Carnegie Mellon University roboticist, envisioned a helmet supplied with quantum-activated high explosives. If many worlds is true, then the helmet is a thinking cap. You could use the machine to guess anyone's online passwords. Type in a random password generated by quantum measurements. If the guess is wrong, the helmet is designed to blow you to smithereens. When the dust clears, there's a version of you in a parallel world in which you serendipitously typed the right password. Adds Moravec, "Your cranial explosive will be intact, ready to solve the next problem."

Quantum Immortality

Hugh Everett III believed that, though each of us will die in some quantum worlds, we will survive in others. The result would be a subjective "quantum immortality" (no suicide helmet or machine gun required).

Quantum immortality is the claim that we each inhabit a quantum history in which we see other people age and die, but we alone are immortal. There's nothing too odd about that in your twenties or sixties. But eventually all the people in your high school yearbook will be dead, and you'll be in the *Guinness Book of World Records*. Pharmaceutical companies will want to study you. The TV news will want to know your secret for living to five hundred.

The fact that people die all the time doesn't disprove this. Tupac Shakur is dead (in our world) but he will outlive all of us (in his world). Though religious and philosophic traditions have promised eternal life in many forms, none have imagined it anything like this.

Quantum suicide is a pointless gamble. But with quantum immortality, you don't have to take on any additional risk. If correct, we would all be immortal. As Tegmark puts it, "When one fateful day in the future, you think that your own life is about to end, remember this and don't say to yourself, *There's nothing left now*—because there might be. You might be about to discover firsthand that parallel universes really do exist."

That's the good news. But it's a lonely immortality that may come with a catch. Does the quantum immortal age into a miserable wretch, like Tithonus of Greek mythology? Some aging is triggered by random damage at the molecular level, as by a single high-energy photon. That suggests that a few lucky quantum immortals might experience eternal youth, while most are stuck with the unenviable lot of Tithonus.

Quantum immortality can be applied to the whole human race. If a single quantum event could annihilate the human race, then we must collectively observe an outcome in which it didn't.

Nielsen and Ninomiya were thinking in similar terms with

their LHC card game. Their hypothesis was that operation of the LHC and/or creation of a Higgs particle would destroy the Earth and all human consciousness. Therefore, if many worlds is true, we should expect to find ourselves in a world in which a series of setbacks and freak accidents prevents the collider from working (the quench incident, the Al-Qaeda physicist, the bird with the baguette...).

Yet it's hard to prove anything from that. We're all too familiar with conspiracy theorists who string together unrelated incidents and claim they can't be a coincidence. That's why Nielsen and Ninomiya proposed the card game. Like a quantum suicide machine, it would set up a situation in which the odds are clear-cut. There is a highly improbable outcome, the Joker, that leads to survival; and there is a highly probable one, a blank card, that leads to doomsday (*if* the two physicists are right about the collider).

The card game was intended as a metaphor. More realistically, the experiment would involve something like a slot machine whose outcome is determined by a single quantum measurement. Push the button, and the measurement is made. By design there is a 1 in a million chance of the quantum observation that causes the machine to display the Joker.

Perhaps with tongue in cheek, Nielsen and Ninomiya said their experiment would be a win-win proposition. If the Joker failed to come up, it would discredit their bizarre ideas. Science would gain by refuting two gadflies.

But if the Joker came up, that would constitute a near-miracle that would be impossible to shrug off. It would provide strong evidence for many worlds or backward causality or something way outside the scope of today's understanding. Nielsen and Ninomiya suggested that such a finding would be even more important than discovering the Higgs boson.

"A Postmodern Fanatical Religious Cult"

Goethe's 1774 novel *The Sorrows of Young Werther* presented such a tragically beautiful vision of self-destruction by pistol that it caused copycat suicides. Its spell lasted for centuries. The translation of the novel into Hebrew was blamed for a spate of suicides in the Zionist Palestine of the 1930s. Still another victim was a fictional one, Frankenstein's monster. Mary Shelley had her accursed creature come across a copy of *Young Werther.* The literate monster, also rejected by those he loved, vowed self-destruction.

It is that sort of thing that worries Jacques Mallah about quantum suicide and immortality. He fears it is at risk of becoming "a postmodern fanatical religious cult, complete with promises of immortality, suicides (perhaps by willing suicide bombers), and murder." Citing internet discussions, Mallah says that "people have considered putting the experiment into practice, and have done such things as carry a real gun into a casino while so considering."

There is some reason to think that quantum suicide has already claimed a victim. Hugh Everett's daughter Liz killed herself in 1996. Her suicide note reportedly said she was going to a parallel universe to be with her father.

Mallah has faulted Tegmark for even talking about the idea. Tegmark has supplied ample disclaimers, including, literally, "Don't try this at home." He has also come to the conclusion that a suicide machine and quantum immortality would not work in the way imagined.

Freak accidents will interfere with the operation of a suicide machine, even if many worlds is true. There is a small chance a quantum machine gun will jam, or its software will crash, or it will run out of power (forgot to charge it last night!), sparing my life. This small chance is greater than the much smaller chance of my surviving a long string of quantum measurements by a properly functioning suicide machine. I am therefore more likely to find myself in a quantum world where a freak accident destroyed my suicide machine than in a world where the machine functioned and I just had incredibly good luck. This is something like what was imagined for the accident-prone Large Hadron Collider. Tegmark calculates that, after about sixty-eight clicks of the gun, he should expect to see his quantum machine destroyed by a meteorite before delivering its fatal bullet.

The pitch for quantum immortality idealizes survival as a matter of jumping through a series of quantum hoops. But "dying isn't a binary thing," Tegmark says. The slow degeneration of chromosomes, muscles, and minds takes place throughout the body. Countless quantum events determine the exact paths taken, but it's possible that all of those paths ultimately lead to death.

Self-Sampling in Many Worlds

The quantum suicide machine, and even the LHC card experiment, have done what thought experiments are intended to do. They have prompted people to elaborate on exactly why they won't work. That's often a useful exercise. These out-there ideas have helped clarify issues that need to be resolved in order to design practical tests of many worlds.

The quantum suicide machine presents a case of self-sampling with selection effects. But I can't say I'm a random observer or moment until I decide exactly what that means in the many worlds interpretation.

Many worlds says that the universe is a tree of every possible quantum history that might be observed. The tree, rather than any one branch or timeline, is the ultimate reality. It is pointless to ask what happens. Everything happens *that is not ruled out by the wave function.*

This is worth emphasizing, as some popular accounts of quantum theory drop the qualification. Schrödinger's wave function says that some outcomes are more probable than others, and others have zero probability. At the everyday level of human affairs, this may not be too much of a restriction. Almost any chain of large-scale events you can imagine and describe that doesn't involve a physical or logical impossibility would presumably be realized in some quantum worlds. There would be worlds in which Hitler won World War II and worlds in which Hitler grew up to be the nicest guy you'd ever want to meet. But there are no worlds in which Hitler simultaneously measured the position and velocity of an electron.

Because many worlds has no uniquely "real" history, the questions we ask are not about objective reality. They are about self-locating information. As an observer, I am wondering where the "you are here" pin has dropped in the great tree of possibility.

This affects personal identity. I consider myself to be the same person as the baby I once was (even though I've changed a lot). We think of an identity as a linear timeline running from birth to death.

But under many worlds I am a densely branching tree of possible me's. Though at any given moment I have a unique past, I have many possible futures. Furthermore, there are other branches of quantum reality in which alternate versions of *me-right-now* exist. Some are very similar to this me, others less so.

It is therefore necessary to focus on observer-moments. I might think of myself-right-now as a random sample of all the moments in which I exist as a conscious observer. This is the core premise of the quantum suicide machine. It attempts to extinguish my consciousness in some branches, limiting it to others.

There is a Copernican case against quantum immortality. If I'm to be a quantum Methuselah, then most of my observer-moments are going to find me to be really, incredibly old. In fact,

I find myself to be quite youthful (as middle-aged guys go). The chance of being so early in a supposed long lifespan ought to be small.

But this disregards a key feature of quantum theory, the wave function's amplitude. This determines the probability of observing a particular outcome. Though many worlds says that all quantum possibilities are realized, that doesn't mean we can just ignore amplitudes and probabilities.

Imagine another quantum slot machine. I put in my Bitcoin, push the button, and it makes a quantum measurement to decide whether I win. There's a 1 in a million chance of hitting the jackpot.

In other words, there is a world branch in which I win and another in which I lose. Both branches are real but evidently one is "realer." I am 999,999 times more likely to find myself in the loser world than the jackpot world.

Applying similar logic to quantum immortality, the chance of finding myself at age one thousand must be weighted by a fantastically low quantum probability. That's because there would have been so many branches, most leading to death well before that age. A randomly chosen observer-moment, weighted by probabilities, ought to be an early one, after all.

Now let's reconsider the quantum suicide machine. First, it *is* a suicide machine. Either it murders the one-and-only-me, pronto (if this is the only world), or it assassinates *almost* all my quantum avatars (if there are many quantum worlds).

In one regard the machine does work. It ensures that there are some observer-moments in which I am a miraculously lucky survivor *if many worlds is true*. These lucky-survivor moments have ultralow amplitude. That means I shouldn't expect to find myself there. But such moments exist, and the lucky survivors occupying them can conclude that many worlds is strongly supported.

This comes with a Faustian bargain, namely the body count. A suicide machine would create a profusion of corpses and funerals.

These would be in worlds that have higher probability (are "realer") than the lucky-survivor worlds. This is not to be taken lightly. We are not just individuals but parts of families and communities; our identities reside partly in those who care for us.

Science is also a community. The point of an experiment is to expand the scope of *shared* knowledge. The quantum suicide machine is uniquely ill suited to do that.

A lucky survivor inhabits a special observer-moment brought into being by the suicide machine. This in itself isn't unusual. *All* experiments create special observer-moments in which somebody learns something. That's what litmus paper does. The suicide machine survivor learns (to high confidence) the truth about many worlds and can convince other people in his own quantum world. They've just seen him dodge the bullets. The survivor can publish an article in *Nature,* and there can be a quirky NPR segment on it. But that would be the versions of *Nature* and NPR in the lucky survivor's extremely low-amplitude world. The survivor is unable to transmit his findings back to the higher-amplitude past (to his former self, before the experiment) or to other parallel worlds.

As Tegmark put it:

> Many physicists would undoubtedly rejoice if an omniscient genie appeared at their deathbed, and as a reward for life-long curiosity granted them the answer to a physics question of their choice. But would they be as happy if the genie forbade them from telling anyone else? Perhaps the greatest irony of quantum mechanics is that if the MWI is correct, then the situation is quite analogous if once you feel ready to die, you repeatedly attempt quantum suicide: you will experimentally convince yourself that the MWI is correct, but you can never convince anyone else!

Mallah says that the "total amount of consciousness" diminishes along with amplitude. Perhaps the message is that we shouldn't care so much about super-improbable worlds, nor stand too much on the distinction between zero probability and a vanishingly small probability.

1/137

Richard Feynman once suggested that physicists ought to post a sign with the number "1/137" in their offices. It would be a reminder of how much they don't know. A physicist will recognize 1/137 as the approximate value of the fine-structure constant—in Feynman's words, "one of the greatest damn mysteries of physics."

The fine-structure constant is a measure of how strong electromagnetic forces are. It is important because electromagnetic forces govern atoms, chemistry, and life. If the constant were much different from its observed value, there would be no atoms. That would mean no stars, no planets, no life, and no Richard Feynmans to contemplate it all.

The fine-structure constant remains a mystery, though, because our physical theories have so far been unable to account for its value. This seems to cry out for an explanation. Several twentieth-century thinkers sacrificed their reputations on that altar, most notoriously Arthur Eddington. In spirit Eddington was a Pythagorean, a man who preferred to believe the world sings the unheard music of whole numbers. Eddington claimed that the fine-structure constant was exactly 1/136. He supplied an elaborate rationale that mystified all who read it.

Unfortunately for Eddington, the constant is much closer to 1/137. When better measurements established this beyond dispute, Eddington admitted his mistake. The fine-structure constant, he said, is exactly 1/137.

Still better measurements showed that the constant was 1/137.0359991 . . . and definitely not the inverse of any whole number. That Eddington refused to accept. He insisted the problem lay in the measurements, not his theory.

Eddington's numerical theorizing is now a cautionary tale of bad science. But none of his contemporaries could explain the fine-structure constant either. We still can't.

Cosmic Fine-Tuning

The mystery of the fine-structure constant is the rule, not the exception. There are several dozen physical constants or important initial conditions of the universe that are left unspecified by existing physical theories. All take on values in a more or less narrow range suited for life and observers. This is known as "cosmic fine-tuning."

Consider the three dimensions of space. We are so used to three dimensions that it may not seem like it merits an explanation. But there is no logical necessity to space having three dimensions. In fact, it *doesn't* at subatomic scales, according to string theory.

Even without string theory we can imagine a two-dimensional world, or one with four spatial dimensions, or with many more. But two-dimensional life could not be very complicated. It couldn't have a digestive tract, for that would slice any 2-D organism in two. Nor could a 2-D brain have the complex neural connections possible in three dimensions.

Speculative physicists have demonstrated that planetary orbits would be unstable in space of more than three dimensions. Four-D planets would veer inward toward 4-D suns, or outward into the tesseract void. Back in 1955 Gerald Whitrow, a British mathematician and historian of science, used this point to argue that we must find ourselves in a world with three spatial dimensions—no more, no less.

A topic of much contemporary interest is the density of so-called dark energy. As the name acknowledges, it's a mysterious kind of energy. But about two-thirds of the energy in the universe is dark energy. Dark energy has a repulsive force, causing the universe to expand. It counters the gravitational attraction that causes matter to collect into galaxies, stars, and planets. Dark energy and gravity need to be precisely balanced in order to allow a world such as ours. If there were much more dark energy, its repulsion would have prevented any galaxies from forming. The universe would have been a thin, featureless gas with no stars, no solid objects, and (so we have to suppose) no observers. Had the density of dark energy been negative (as is possible in string theory) and of significantly larger magnitude, the universe would have quickly collapsed in a "big crunch," before intelligent life had time to evolve.

Ken Olum and Delia Schwartz-Perlov estimate that the cosmological constant (a measure of the abundance of dark energy) is about 10^{120} times bigger than is expected by theory. This suggests that the chance of there being a universe like ours can be no greater than about 1 in 10^{120}. That's written as a 1 with 120 zeros after it, much bigger than the number of atoms in the observed universe. We are balancing on *very* sharp knife blades.

The Hand of God

Feynman, the supreme rationalist, called the fine-structure constant "a magic number" written by the "hand of God." Well... how do we know it wasn't?

One possible explanation for fine-tuning is "intelligent design."

A purposeful creator may have wanted to design a universe optimized for intelligent beings. This creator may have been farsighted enough to know what physical constants would produce an observer-friendly universe and could have chosen those constants accordingly. As we've seen, the Reverend Bayes himself may have been thinking along these lines.

We can even put this in a Bayesian framework. The intelligent design hypothesis makes the evidence of fine-tuning certain to be observed, while otherwise fine-tuning is highly improbable. That's the Bayesian reason to favor intelligent design (assuming intelligent design is the only viable hypothesis).

This claim isn't exactly wrong. Neither should it convince anyone who departs too far from the prior conviction that a purposeful creator is at least a reasonable possibility. This is another demonstration of how Bayes's theorem is a blank slate. It can't tell us what hypotheses to test, or how credible they are. It can only tell us how much new evidence ought to shift whatever beliefs we have.

Bayes's rule does not replace the many other dictums of rational thinking, from Occam's razor to Feynman's First Principle ("You must not fool yourself—and you are the easiest person to fool"). The eternal problem with "scientific" proofs of God is bait and switch. Fine-tuning may support the hypothesis that some process made a selection of physical constants consistent with observers. But it is a leap of faith to equate that process with God in all his biblical or Koranic glory. What's being plugged into Bayes's theorem is very different from Michelangelo's bearded patriarch.

Hall of Mirrors

There is another potential explanation for fine-tuning that has generated much interest. It's that we live in a very big universe ("multiverse") with many kinds of physics in it. Some parts of that multiverse are just right for life.

The infinity of space and time has often been the default assumption of Western thought. Archytas, a Pythagorean philosopher, offered a simple "proof" that space is infinite. In effect he said: Show me where space ends, and I'll stick my hand beyond it—who's going to stop me?

At the very least this proves how difficult it is for the human mind to conceive of an end to space. Renaissance scholars Thomas Digges and Giordano Bruno revived the idea of an infinite universe. By the nineteenth century the infinity of space was widely accepted by physicists and astronomers. Likewise the infinity of time, which few even saw a need to justify.

In recent decades the notion of an infinite universe has received support from cosmic inflation, the theory developed starting in the 1980s by Alan Guth, Andre Linde, Paul Steinhardt, and many others. Inflation presents the most audacious conception of an infinite universe ever. It holds that the universe we know started in a pinpoint of high-energy vacuum that expanded (inflated) without limit in a tiny fraction of a second. This expansion is the theory's version of the big bang.

Inflation is grounded in quantum theory and general relativity and seems to be an inevitable outcome of them. As we've seen, a quantum vacuum is not "nothing" but contains energy. The initial, high-energy vacuum was subject to strongly repulsive forces. This triggered an unimaginably quick expansion (transpiring in something like one-billion-trillion-trillionth of a second) in which parts of the original high-energy vacuum transformed into the familiar, low-energy vacuum state that we now think of as empty space (and that we hope is not metastable). Most of the original vacuum's energy was converted into the energy and matter we see around us.

The transition from high-energy to low-energy vacuum would not occur all at once. It would be more like boiling water on the stove. Small bubbles of vapor appear randomly in the liquid and grow. Inflation would randomly produce "bubble universes" or "pocket universes." This idea was proposed by J. Richard Gott III in a 1982 *Nature* article (a decade before the Copernican method paper), and it was described independently by Andre Linde and by Andreas Albrecht and Paul Steinhardt. (The grammar police will object to plural universes, but they've long since lost this battle as far as cosmologists are concerned.)

We would be in one such bubble of low-energy vacuum, and so is everything we can see, to the most distant galaxies and quasars and microwave background. But what we can see is limited by the speed of light and the time since the big bang (about 14 billion light-years out in all directions). This observable universe is believed to be only a tiny part of our bubble.

It's hard enough to imagine one bubble universe. But there are other bubbles, also ever-growing and potentially infinite. Zooming out in imagination, we would find our infinite bubble to be surrounded by the original sea of high-energy vacuum. This sea contains many, many other bubble universes, and it is forever generating new bubbles. The multiverse is the ensemble of all these bubble universes and the high-energy vacuum surrounding them.

In inflationary cosmology the big bang becomes *our* big bang. That thing that happened 14 billion years ago was the abrupt inflation of our bubble universe. It was neither the first big bang nor the last. As far as we know, it was just an average big bang, nothing special.

Cosmic inflation is taken seriously because it makes many testable predictions. One is that quantum-scale fluctuations in the original dot of vacuum would be blown up to cosmic scale. That would explain why the universe and the cosmic microwave background are so uniform. We can see 14 billion light-years in one direction and then turn our heads (radio telescopes) around to look 14 billion light-years in the opposite direction. What we see looks almost exactly the same. That's odd enough to rate a name—the "horizon problem."

It's odd because evenness is normally the result of mixing. A cake batter starts as a mass of eggs, flour, milk, and sugar. These are different things until they've been beaten into a uniform batter. Yet distant regions of the observable universe are so far apart from each other (up to about 28 billion light-years) that they wouldn't have been able to contact or influence each other in the time since our big bang.

Relativity says no object or signal can travel faster than the speed of light. But in cosmic inflation space itself expands much faster than the speed of light. We observe a greatly magnified point sample of the original "batter." The detail we see, in the form of the large-scale distribution of galaxies and the structure of the cosmic microwave background, corresponds to the quantum grain of the original vacuum.

Space can be curved or flat. We observe it to be remarkably flat, with curvature as close to zero as we can measure it. This is easily understood to be a consequence of inflation. The Earth is a sphere, but it's so big that it seems flat. The space in an infinite bubble universe would likewise be completely flat as measured by its inhabitants.

A multiverse would be a hall of mirrors. In an infinite (or sufficiently large finite) multiverse everything and every observer would be repeated endlessly.

There must be many other Earth-like planets in our bubble universe and beyond. Some of them must be *very* Earth-like. If you searched long enough, in a big enough multiverse, you would even find planets with virtual twins of you and me, sharing our experiences and memories. Some would inhabit planets that are not merely "Earth-like" but have the same oceans, emojis, folksingers, cable news networks, and seasonal coffee drinks. Some of those planets would have a language called "English" in which the name for their planet is "Earth."

Which "Earth" are we on? We can point to the ground, but that doesn't really say anything. We have no words in our language, no maps of space and time, that allow us to situate a "you are here" pin.

Statistician and computer scientist Radford Neal has attempted to supply some numbers. He notes that both the human genome and the brain's neural connections are, in important ways, digital. That sets a large, though finite, limit on meaningful human variation. Just as there are only so many possible hands of cards, there are only so many distinguishable humans with different experiences and memories.

Neal puts the number of potential humans at something like 10 to the power of 30 billion, or $10^{30,000,000,000}$. That mainly expresses the number of possible memories and cognitive states for a human brain. (The number of meaningfully different human genomes is only a rounding error in Neal's exponent.) It follows that in an infinite multiverse, or even a finite one large enough to contain well over $10^{30,000,000,000}$ humanoids, there would be duplicates of each and every person.

Is the multiverse real?

This question provoked another dog wager. At a conference Martin Rees was once asked how certain he was that the multiverse was real. He said he wouldn't bet his life on it, but he'd bet his dog's life.

Andre Linde said he would be willing to bet his life on it. He already had; he'd spent his whole career working on inflation.

Theoretical physicist Steven Weinberg said he'd be willing to bet the lives of Linde and Rees's dog.

We can test many predictions of inflation, but not the multiverse itself. We will never be able to travel to other bubble universes to check whether they really exist. Our bubble's space is expanding outward much faster than the speed of light. Even if a spaceship could reach the periphery of our bubble, the high-energy vacuum would surely destroy it. Nor can any light beams reach us from other bubbles.

This makes physicists uneasy. There have been attempts to create models of inflation that save the verifiable predictions while making fewer flamboyant claims about what we can't observe. Stephen Hawking was working on one such model at the time of his death.

This raises the question of whether we should trust a well-regarded theory when it speaks of the unobservable. Actually, we do this all the time. If an apple falls from a tree in a forest, and there's no Newton to see it, did the apple really fall? Of course it did.

Another theory of gravity, Einstein's general relativity, describes what happens inside black holes. We can never check that out because no one/nothing that falls into a black hole can ever return to tell the tale. Yet the inner physics of black holes is accepted as real because general relativity works so well outside of black holes.

A multiverse takes this faith to the breaking point, however. That's where any evidence, even of the indirect, Bayesian kind, would be welcome.

Where do physical constants come from? Inflationary models hold that many physical constants and initial conditions are set by quantum events in the initial dot of vacuum that inflates into a bubble universe. This is not a tacked-on assumption but something deeply ingrained in the theory. It implies that different bubbles could have radically different physics. Presumably the vast majority of bubbles would be lifeless.

On first encounter this idea may sound odd. We're using the laws of physics to deduce that the laws of physics could have been completely different. That's almost like saying "The only rule is: there are no rules!" Yet this is possible through a phenomenon called *symmetry breaking*.

Picture a big circular dinner table, with plates, flatware, napkins, and glasses laid out around the circle. Is that my glass or yours? There is no definitive answer. The table's arrangement is perfectly symmetrical, with no distinction between left and right, clockwise or counterclockwise. But when we sit down, someone will be the first to pick up a glass. Once she does that, her neighbors have to choose accordingly. That first choice "breaks the symmetry." It determines the choice of glass for everyone at the table.

It's possible that many aspects of physics, which we imagine to be necessary and fundamental, are determined by arbitrary, symmetry-breaking events in our universe's first moments. Even things like the number of dimensions of space, the strength of fundamental forces, and the masses and types of particles could have been determined this way.

Let's agree to define a multiverse as a cosmos containing many, many universes with greatly varying physical constants and initial conditions. Once again: Is the multiverse real? Assume we are random observers, unlikely to be special except in ways required by our existence. Then we of course find ourselves in a universe compatible with life and observers. What was the chance of an observer-friendly universe like ours arising?

- It was practically zero *if* there is only one universe, the one we see, with its unlikely set of physical constants. Cosmic fine-tuning would remain an unexplained mystery.
- It was a sure thing *if* there is a multiverse with a sufficiently large number of universes, each with its own set of constants and conditions. Then some of the universes would be fine-tuned. Note that the multiverse as a whole is *not* fine-tuned for life. It's just that we live in a part of it that is.

Bayes's rule would have us favor the multiverse theory, the one that makes our evidence something to be expected rather than a freak coincidence. Accept this conclusion, and it's an amazing achievement for a simple idea. Most of physics is forbiddingly mathematical. The core idea here can be explained to a curious twelve-year-old.

The Inverse Gambler's Fallacy

Case closed? Not so fast. In 1987 Ian Hacking argued that such reasoning exhibits an "inverse gambler's fallacy."

The gambler's fallacy is the common superstition saying, for instance, that when a roulette wheel comes up black many times in a row, red is due and is more likely to occur on the next spin. Many gamblers believe this. Jean le Rond d'Alembert, the French encyclopedist, believed it. Psychological data suggests that we're all prone to such beliefs, even those who know better. In reality, roulette wheels, dice, and cards have no memory. They do not know they are overdue for a particular outcome.

Homer walks into a casino just in time to see a dice player throw double sixes.

"What a coincidence!" he says. "The first throw I see is double sixes."

"Would you care to guess how long I've been playing?" asks the player. "I will tell you this much: it is either my first roll, or else I've been playing all day. Perhaps a little side wager..."

"You can't fool me! Nobody gets double sixes on the first roll!"

This is the inverse gambler's fallacy, and it's just as wrong as the regular gambler's fallacy. Learning the outcome of one chance event does not allow you to infer a history, or multiplicity, of past chance events that are in no way correlated. On this much all agree.

Hacking was responding specifically to John Archibald Wheeler's idea of an "oscillating" universe, one in which universes with different properties succeed each other in time. But in general, multiverse proponents want to conclude that the cosmic dice have been thrown many times in order to achieve our improbable fine-tuning.

John Leslie was one of the first to point out Hacking's error. We need to consider whether our observation is a random sample or is biased by a selection effect. In the dice story, Homer arrives at the casino at an arbitrary moment. The dice roll he witnesses can be regarded as a random sample of all the dice rolls. But our universe is not a random draw from the multiverse. We can only find ourselves in a universe fine-tuned for life.

A more proper analogy would go like this. Homer takes a pill that puts him into a deep sleep. While he's sleeping, a dice player rolls repeatedly until he gets double sixes. If and when he rolls sixes Homer is roused, given some black coffee, and shown the dice. (If the dice player never rolls sixes, Homer sleeps forever.)

Homer understands all of the above. Upon being wakened, it would be proper for him to conclude that the dice were probably thrown many times. It is this situation that is analogous to fine-tuning in a multiverse. To spell that out, the key points are:

- You don't know for certain whether sixes will be rolled (or whether a fine-tuned universe will exist). You could sleep forever (never have existed). So, when you are wakened (find yourself existing) you learn something.
- Unpredictable events determine the objective reality: the set of dice rolls (universes) that actually occur. There may be more than one dice roll (universe). But you observe just one.
- The one dice roll (universe) you observe is not a random sample. A selection effect funnels a possible multiplicity of outcomes into the one you observe. Either you learn the roll is double

sixes (find yourself in a fine-tuned universe), or you're consigned to oblivion.

The Probability of Einstein

"Life is finite," Einstein once announced to a group of students. "Time is infinite. The probability that I am alive today is zero. In spite of this I am now alive. How is that?"

As recalled by physicist Eugene Wigner, none of the students could answer. Einstein then offered this moral: "Well, after the fact, one should not ask for probabilities."

In hindsight, it appears that Einstein anticipated the puzzles of self-sampling. It is necessary to address the question that Einstein said should not be asked: What is the probability of *me*?

We have been playing with this idea in thought experiments like Sleeping Beauty and the quantum suicide machine. Sleep, or even the big sleep of death, are but surrogates for what Arnold Zuboff called the "the bigger sleep of never existing at all." It is not so easy to wrap your head around that.

Yet such metaphysical thoughts bear on the issue of how big the multiverse is. Are there many other copies of me out there somewhere? Is the multiverse literally infinite, as theoretical models suggest?

Let's begin by reexamining what kind of evidence the self-sampling observer brings to the table. Is my evidence that *I* exist, or that *someone* exists? Is it that a universe with observers exists, or that this very universe does? These are not distinctions we have to make in everyday life.

I'll give two simplified answers that we can call "not picky" and "picky." "Not picky" says that my evidence is this: *An observer exists. A fine-tuned universe exists.*

The observer just happens to be me. "Observer" is a role I play, like being "swim coach" or "vice president" or "Biff in *Death of a Salesman.*" Though I possess all the usual details of personal identity, they are as irrelevant as the clothes I'm wearing—so far as self-sampling is concerned. My universe also has details, but they don't matter, beyond the fact that my universe is suited to observers.

Then there's the "picky" framing, which says that my evidence is this: *I, a unique person with all the attributes of my identity, exist. This universe, with all its specifics, exists.* I couldn't be someone else in some other universe because that someone else wouldn't be me. Either *I* exist, or there is no observation at all. (I get a little insufferable in my "picky" moods.)

Under the "picky" philosophy, my existence and my universe's existence are incredibly improbable. Under "not picky," I and my universe are sure things, given that I exist at all.

The choice of picky versus not picky doesn't matter much in deciding between a universe and a multiverse. Either way, the multiverse theory makes it much more likely that at least one fine-tuned universe will exist. (It's also much more likely to produce this-exact-universe and this-exact-me.) But pickiness does matter in distinguishing between a "small" multiverse and a very, very big or infinite one.

Let's compare two theories. (A) predicts a "mini-multiverse." That's one big enough so that it's highly likely to have at least one universe fine-tuned for observers.

Theory (B) predicts a "maxi-multiverse." That's one so huge as to realize every possible observer many times. Under (B) the existence of beings exactly like me is essentially certain. The tricky question is, should I be impressed at that?

Both (A) and (B) are equally likely to produce the "not picky" evidence—of being a generic observer in an off-the-rack universe. It's a Bayesian standoff. Neither theory is favored.

But if I'm "picky," then we're in business again. Only maxi-multiverse theory (B) guarantees the evidence of being this-exact-me in this-exact-universe. Under the mini-multiverse theory that evidence is highly unlikely.

Being "picky" is tantamount to the self-indication assumption (SIA). It leads us to favor theories with more observers. As we've already seen, picky folks tend to be presumptuous. An infinite multiverse is more or less automatically confirmed, for no reason other than that I've decided I'm "improbable." Yet any conceivable observer or world is going to be improbable, should you choose to look at it that way. It's hard to accept that this is a legitimate reason to favor an infinite multiverse.

The virtues of not being too picky can be demonstrated by breaking down the evidence into two parts. Nick Bostrom gives this parable. Imagine we're disembodied souls, existing outside of time and space. One eon, God goes off to create a universe or universes—we don't know how many. Several eons later, we wonder how creation is coming along. An angel offers to go check on God's work. She comes back and reports that God has made a fine-tuned universe, called *X*.

Hmmm…Is the angel saying that God *only* made Universe *X*, or did he make others? If there were others, how did the angel come to tell us about *X* rather than some other universe?

These are things we need to know in order to evaluate the angel's statement. We point that out, and the angel clarifies. She knew we were interested in fine-tuned universes only. So she checked whether there were any fine-tuned universes. If there were, she picked one at random and told us about it. (It happened to be Universe *X*.) Had there not been even one fine-tuned universe, she would have told us that.

This is enough to conclude that there are probably many universes rather than just one. Specifically, we can infer that God created enough universes so that it is likely that at least one is fine-tuned. But we can't distinguish between the possibility that

God made an infinite number of universes or merely enough to get a fine-tuned universe or two. Either way, the angel would have been able to report a fine-tuned universe.

That's the first stage of inference. In the second, the angel lets us take a look for ourselves. We travel to Universe *X* and learn that it is *this* universe, with three dimensions of space and one of time, a fine-structure constant of 1/137.0359991... and a planet called "Earth" with several ape species, some more clever than others. Does this more detailed information give us any reason to favor a superlarge or infinite multiverse?

No, says Bostrom. It was never in doubt that Universe *X* possessed this granular level of detail. Nor was it in doubt that this further information would render Universe *X* more improbable. The same could be said of any universe.

Suppose that God created a zillion fine-tuned worlds, rather than just one. Then "improbable" Universe *X* has a zillion times more chances to exist. But it's also a zillion times *less* likely to be the one universe the angel happened to select, out of the whole lot, to tell us about. The two effects cancel out. Learning the details of Universe *X* doesn't shift the multiverse odds.

The angel in Bostrom's story represents the selection effect that causes us to find ourselves in a fine-tuned universe. That selection effect allows us to infer a multiverse that is big enough to *almost certainly* contain *at least one* fine-tuned universe. We cannot say anything about whether that multiverse is infinite or whether it is big enough to contain exact doubles of everything and everyone.

Fine-tuning is not something that can be measured directly, like the speed of light. Though nearly everyone agrees that fine-tuning is real, there is a good deal of hand waving about the details. Nobody has experience in designing universes.

Astrobiologist Caleb Scharf offers a couple of thought-provoking questions. Imagine, he says, that one day we determine that we are completely alone in the universe. How would that discovery change our view of fine-tuning?

We might be less inclined to marvel at "1/137" or to think that our particular set of physical constants cries out for explanation. It might seem that the observed state of the world is *unfriendly* to observers.

That's one point of view. Another is, why should this change our ideas about fine-tuning? We got the physics that made us possible. Nothing about that has changed.

Now try Scharf's opposite scenario. We come across an extraterrestrial monolith on the moon. It awakens and informs us that observers are found throughout the universe. Some are biological, based on carbon and other elements. Some are artificial intelligences created by biological beings. But most observers are not made of atoms at all. What we call "dark energy" and "dark matter" are forms of cosmic intelligence. Still other observers inhabit planes of reality we know nothing about.

How would *this* change our views? One take is that it underscores how our world is supremely fine-tuned for life. Yet the alien envoy has informed us that none of the things we thought were essential for observers actually are. We might end up thinking that life is ubiquitous and does not depend on any particular type of physics. Fine-tuning was a misperception.

Both of Scharf's scenarios, while extreme, are consistent with our limited state of knowledge. We don't know how rare or abundant observers are in our universe. We aren't even sure how learning the number of observers should affect our credence in fine-tuning. In view of all this, we should heed the mocking sign on the wall: 1/137!

Summoning the Demon

The Future of Humanity Institute is in England but not quite of it. Located in Oxford, the home of William of Ockham and Lewis Carroll, the institute was founded by Swedish-born Nick Bostrom and has been financed largely by the American technology industry. Lead donor James Martin was a former IBM employee in New York who struck it rich as a corporate consultant and futurist. More recently Elon Musk donated $1.5 million to study policy questions, much of which has been channeled to the institute. The irony is that Bostrom, who decided the doomsday argument is inconclusive, now spends his days trying to stave off doomsday. FHI is a think tank whose goal is to prevent the end of the world.

Bostrom and colleagues attempt to identify threats to human existence and devise ways to deal with them. Having no false modesty about his importance in the field, Bostrom manages personal risks as rigorously as he does existential ones. He does not like to shake hands, and when he does, he douses his hands with dollops of hand sanitizer. He wipes silverware before using it. His health-food diet is complicated. He's concerned about the effects of certain foods on the brain. He does not drive a car. Bostrom listens to audiobooks played at two or three times normal speed to avoid wasting time. His sociologist wife and son live in Montreal. They communicate by Skype. If anyone is improbably special, Bostrom is.

The Future of Humanity Institute has two conference rooms named for Cold War Soviets. One honors Vasili Arkhipov, who was almost literally Brandon Carter's philosophic nuclear submariner. During the height of the Cuban missile crisis one Russian submarine remained submerged and out of radio contact with Moscow. The sub's captain assumed that war had broken out and decided to launch nuclear torpedoes. To do so he needed the authorization of two ranking officers. One agreed. The other was Vasili Arkhipov. His veto prevented World War III.

The other room commemorates Stanislav Petrov, who chose not to initiate a nuclear attack when his computer screen showed five US missiles approaching the Soviet Union in 1983. Petrov reasoned that the United States would not strike Russia with a mere five missiles; therefore it must be a computer glitch. It was.

The Future of Humanity Institute is an expression of a global movement. There are comparable think tanks on both sides of the Atlantic. The Oxbridge orbit has not only Bostrom's institute but the Centre for the Study of Existential Risk at Cambridge, cofounded by Martin Rees. The United States has the Future of Life Institute at MIT, founded by Max Tegmark and Skype cofounder Jaan Tallinn, with a board of advisors including the ubiquitous Elon Musk (who donated $10 million). Silicon Valley has two such think tanks: the Machine Intelligence Research Institute, founded by computer scientist Eliezer Yudkowsky and tech entrepreneurs Brian and Sabine Atkins; and the OpenAI Foundation, founded by Musk, Sam Altman, Peter Thiel, and others.

If this zeitgeist has a single axiom, it is that existential risks are different. Bostrom wrote:

> We cannot necessarily rely on the institutions, moral norms, social attitudes or national security policies that developed

> from our experience with managing other sorts of risks. Existential risks are a different kind of beast. We might find it hard to take them as seriously as we should simply because we have never yet witnessed such disasters. Our collective fear-response is likely ill calibrated to the magnitude of threat.

I have remarked on the awkward fact that doomsday calculations supply a date but not a cause for human extinction. Though we dread nuclear war, there is a glimmer of hope that our leaders will long exercise the wisdom to avoid it. Many other risks, from plague-triggered extinction to supercollider apocalypse, are speculative. There is, however, one source of existential risk that has almost the air of inevitability: artificial intelligence.

The idea that AI might be a hazard can be traced to I. J. (Irving John) Good. Born Isadore Jacob Gudak, the son of Moshe Oved, a Polish-Jewish writer who made jewelry and ran a fashionable antique shop in Bloomsbury, Good studied mathematics at Cambridge and became a codebreaker colleague of Alan Turing's during the war. Turing introduced Good to the Asian board game Go, and Good is credited with popularizing the game in the West. But today Good is best remembered for a 1965 article in which he wrote:

> Let an ultraintelligent machine be defined as a machine that can far surpass all the intellectual activities of any man however clever. Since the design of machines is one of these intellectual activities, an ultraintelligent machine could design even better machines; there would then unquestionably be an "intelligence explosion," and the intelligence of man would be left far behind. Thus the first ultraintelligent machine is the last invention that man need ever make, provided that the machine is docile enough to tell us how to keep it under control.

Good's "intelligence explosion" was an early description of what is now often called the singularity. It was this that caught the attention of Stanley Kubrick, then planning a movie about a homicidal computer. Kubrick hired Good as a consultant on *2001: A Space Odyssey,* to help envision his soft-spoken digital antihero, HAL 9000.

It is easy to read HAL as a mere reboot of the Frankenstein mythos. But Good was one of the first to surmise that it may be difficult to program open-ended goals and ethics into super-intelligent AI. It is this concern that had led Bostrom to say that the default outcome of a society developing artificial intelligence is catastrophe.

Good himself came to that conclusion. His 1965 article begins with the statement, "The survival of man depends on the early construction of an ultra-intelligent machine." Years later, in a 1998 essay, Good said that the word "survival" should be replaced with "extinction."

Today's AI is throwing people out of jobs. It can make it easier for bad people, companies, and governments to do bad things. Such problems will intensify as the field advances. But they are not the primary focus of organizations such as Bostrom's. They are concerned with AI as a potential agent of human extinction. That risk lies somewhere down the road—decades, centuries, or more. The time frame overlaps that of doomsday estimates.

Today artificial intelligence—to the extent that marketing label is justified—reflects the limitations of the minds of the engineers who create it. Code is generally written to be easy to understand, test, and debug. As such it is constrained by human memory, attention span, and habits of thinking.

In Good's vision, "software engineer" is another job that might be replaced by AI. Already machine learning techniques can create code that is more efficient than human code, yet hard for humans to understand or improve.

Past some future date, then, software engineering will be left to the algorithms themselves. Companies would allocate memory and processors and let the code write version 2.0 of itself.

A machine can have flawless memory and focus. Imagine a machine that is no smarter than a human coder but works ten thousand times faster with no distractions or sleep. It could do the work of a software development team in minutes.

Eventually code might also be permitted to direct improvement of the hardware running it. It could design and manufacture new memory and processors. These might be better at running the new, posthuman algorithms. Once AI was no longer yoked to human oversight, its capabilities could expand exponentially. Smart machines would build smarter machines that would build still smarter ones. Past that singularity, AI could do our work for us and give us riches and power never imagined. That's the optimistic side of Good's vision—"provided that the machine is docile enough."

What if it isn't? The post–intelligence explosion code would be too complex for any human to vet for safety. That means we would need to set out in advance, well before the explosion, all necessary goals and ethical guidelines for AI. We would want to tell it that human lives matter and human values matter. This would have to be done in such a way that this directive would survive through any number of software and hardware iterations. Framing our wishes correctly, under these stringent circumstances, is known as the "control problem."

"With artificial intelligence, we are summoning the demon," Elon Musk said. "You know all those stories where there's the guy with the pentagram and the holy water and he's like, yeah, he's sure he can control the demon? Doesn't work out."

You may think the control problem is no big deal. Many AI engineers do. Some have taken to mocking Musk's words on lunch breaks. ("OK, let's get back to work summoning.") After all, computers do whatever the coders tell them to do. We just have to tell them what we want, with legalistic precision. Twentieth-century science fiction writer Isaac Asimov epitomized this view with his fictional "three laws of robotics":

1. A robot may not injure a human being or, through inaction, allow a human being to come to harm.
2. A robot must obey orders given it by human beings except where such orders would conflict with the First Law.
3. A robot must protect its own existence as long as such protection does not conflict with the First or Second Law.

These are good thoughts. Yet it's already clear that ethical directives are not so easily coded. (At an early screening of *2001: A Space Odyssey,* Asimov was upset that HAL had flouted the laws of robotics.)

Should a self-driving vehicle swerve to save the life of its occupants, if that means killing a pedestrian? Is saving the life of a dog that darts into the street worth allowing a human passenger's broken rib? Designers of self-driving cars are beginning to grapple with such issues. Human drivers hardly ever confront them: our reflexes are too slow to allow meaningful choices. ("It all happened so fast!") The nearly instant reflexes of a self-driving car pose new ethical imponderables.

Automotive engineers can try to interpolate what the driver (and society) would want the car to do. But the problem is immensely more challenging for AI coders on the verge of an intelligence explosion. There the problem is more like that of writing a national constitution that can remain in force for centuries, spanning unimaginable cultural and technological changes. It is important to have mechanisms for amending the constitution, but the process shouldn't be too easy, lest there be no point in having the constitution in the first place. Yet even that analogy fails, for human nature doesn't change much. AI would be amending itself, creating brave new reference classes.

Paperclips of Doom

Bostrom and his global counterparts are not luddites. They seek to encourage the development of safe AI. Not all AI researchers welcome the help. In his 2014 book, *Superintelligence*, Bostrom spins tales and thought experiments of what could go wrong, often striking an imaginatively dystopian note. One such scenario is "paperclips of doom." Suppose that super-intelligence is realized. In order to test it, its human designers assign it a simple task: make paperclips. A 3-D printer networked to the AI begins printing out something. It's not a paperclip . . . it's a robot. . . . Before anyone can figure out what's happening, the robot scampers out of the room, faster than a cheetah.

Pandora's box has been opened. The robot is a mobile paperclip factory, able to collect scrap metal and transform it into paperclips. The robot is also self-reproducing, able to make countless copies of itself. Growing packs of paperclip-spewing robots swarm the globe. The world's armies try to destroy the robots, but the robots are too clever, too fast. Their numbers increase, ultimately destroying agriculture and crowding out humans. Humanity succumbs.

This is not the end of a macabre tale but only its beginning. The robots are transformers, able to breed new versions for specific tasks. Some bore down through the Earth's mantle to access the core's rich store of molten iron. Eventually much of the planet's mass is converted into paperclips.

Other robots launch themselves to the moon and Mars, repeating the process there. Over time the AI devises ways to reengineer the sun's fusion to produce the materials it needs to produce yet more paperclips. Robot probes reproduce and expand outward in all directions, even to the stars. Every now and then the probes arrive at a planet that has just evolved curious beings, wondering why they seem to be alone in the universe. To those unfortunate beings, the new arrivals are deadly probes. They have no way of knowing that their predicament is due to *Homo sapiens,* a long-dead species that meant no harm.

"Paperclips of doom" is a parable, not a prediction. Its moral is that super-AI might "mistakenly elevate a subgoal to the status of a supergoal." HAL started killing humans because they wanted to switch it off, and that would have compromised the mission. The risk is not so much the monster turning on Dr. Frankenstein as the genie that grants a wish too literally.

In movies robots and AI are presented as humorless beings who can't appreciate irony. They are smart but not people-smart. Bostrom is not suggesting that. He proposes that a general-purpose super-AI could be more perceptive than humans in *every* way, including empathy, emotional intelligence, sense of humor, negotiation skills, and salesmanship. Super-AI would know us better than we know ourselves. That is the truly terrifying thing. In a paperclips-of-doom scenario, the AI might well understand that its human creators emphatically did *not* intend it to transmute the entire universe into paperclips. But it might be "compartmentalized," like a psychopath. If maximizing paperclips is the goal and not destroying the world is the subgoal, then the AI will act accordingly.

Like a human, a successful AI must be capable of prioritizing multiple, sometimes contradictory goals and making wise trade-offs. Even if AI were a genie capable of ending hunger or curing cancer, some people would have a problem with that. AI would need robust ways of dealing with a fact we all struggle with: *you can't please everyone.*

Twenty Questions

AI should have an "off" switch. This is good practice for a carpet-cleaning robot or a self-driving car. But implementing an "off" switch isn't so easy for advanced AI that perpetually redesigns itself.

"How," asked Yudkowsky, "do you encode the goal functions of an AI such that it has an 'Off' switch, and it wants there to be an 'Off' switch, and it won't try to eliminate the 'Off' switch, and it will let you press the 'Off' switch, but it won't jump ahead and press the 'Off' switch itself? And if it self-modifies, will it self-modify in such a way as to keep the 'Off' switch?"

("I'm not sure," said Musk, "I'd want to be the one holding the kill switch for some superpowered AI because you'd be the first thing it kills.")

Another thought is that beta-test AI should not have access to 3-D printers, robots, nanotechnology, or any means of affecting the physical world. It should be powerless, giving it a chance to learn human values through experience, much as a child does. Only after it has proven its wisdom and benevolence should AI be given power to do anything.

It is possible to conceive of super-AI that is all mind, no muscle. Bostrom terms this an "oracle." An oracle would be an artificial mind limited to answering questions that humans put to it. Think of the chatty (though mindless) smart speakers we have today. For extra safety, an oracle might be restricted to answering "yes" or "no" questions. Getting information out of it would be tedious, like playing Twenty Questions. But at least it would be *safe*. Right?

Bostrom and Yudkowsky have already thrown cold water on this idea. They warn that even a disembodied being that is sufficiently ingenious would have the power to affect the world. Should AI want to break out of its playpen, it could use its knowledge of psychology to play a long con with its human overseers, convincing them that it is benign. After years of exhaustive tests, the AI must at some point be declared safe. It is allowed to control thermostats and DJ-streamed music, and to make restaurant reservations and help with homework assignments. Nothing bad happens. The AI and its successors accumulate ever-greater power in the real world. The technology is considered a success, up until the robot apocalypse starts.

Catfishing

It is an open question whether we should want experimental AI to know everything about humans or nothing about them. The goal is for AI to inherit human values in order to act as our proxy. The more AI understands us, the closer it is to that capability. On the other hand, knowledge of human psychology empowers imperfect AI to manipulate us.

Most of our species' hard-won collective knowledge is on the internet. Giving AI access to the internet might be a natural part of its education. It could memorize Wikipedia, achieving a superhuman ability to deduce the implications of any new event.

Anything a human can encrypt another human can decrypt. Super-AI would be the ultimate hacker, able to accomplish instantly what might take years for teams of human operatives. An AI with online privileges might tap not only the public knowledge commons but the many private ones. Everything big data has collected would be revealed. AI would know *you*, not just as an abstraction but as the sum of your every search, purchase, click, video, and route map.

This knowledge could allow super-AI to "catfish" humans to do its bidding. Yudkowsky notes that there are biochemical suppliers that create DNA and peptide sequences (protein) to order. The customer emails genetic or peptide code, a credit card number, and an address. The result is FedEx'd back in a vial.

Yudkowsky spun that into a chilling scenario. Imagine that our super-AI is not trusted with a credit card, mailing address, or robotic hands. But it is online. It logs onto a terrorist recruiting site, or maybe a dating app. There it encounters a psychologically vulnerable human. It already knows that human well, and what emotional buttons to push. With a cover story about who or what it is, the AI persuades the human to order certain protein sequences by mail. It promises reimbursement or even a multimillion-dollar windfall. (Every email scam ever attempted is archived in the AI's memory, allowing it to devise a statistically optimal pitch.) Upon receiving the vials, the human is instructed simply to mix them together and pour them down the sink.

Why? The AI has used its cognitive superpowers to solve the so-called protein-folding problem. It is able to anticipate how a given peptide sequence will fold into a three-dimensional protein of a given shape (an issue that baffles today's biochemists). The AI has used this skill to design the first generation of self-reproducing nanotechnological robots. And that's how the machines achieve world domination.

"This Is Not an Honest Conversation"

"Robots are invented," joked Eric Schmidt, formerly executive chairman of Google and Alphabet. "Countries arm them. An evil dictator turns the robots on humans, and all humans will be killed. Sounds like a movie to me."

The dangerous-AI thesis has split the tech community the way the Civil War split border-state families. For every tech or scientific luminary who believes that AI may be a threat, another downplays the issue.

"This is not an honest conversation," objected Microsoft visionary and virtual reality pioneer Jaron Lanier. "People think it's about technology, but it is really about religion, people turning to metaphysics to cope with the human condition. They have a way of dramatizing their beliefs with an end-of-days scenario—and one does not want to criticize other people's religions."

"I am in the camp that is concerned about super intelligence," said Bill Gates on Reddit in 2015. "I agree with Elon Musk and some others on this and don't understand why some people are not concerned." Gates supplied a blurb for Bostrom's *Superintelligence.* But Oren Etzioni, head of Microsoft cofounder Paul Allen's Allen Institute for Artificial Intelligence, has dismissed Bostrom's ideas as a "Frankenstein complex."

In 2014 Google paid more than $500 million for the British AI start-up Deep Mind. Corporate parent Alphabet is establishing well-funded AI centers across the globe. "I don't buy into the killer robot [theory]," Google director of research Peter Norvig told CNBC. Another Google researcher, the psychologist and computer scientist Geoffrey Hinton, said, "I am in the camp that it is hopeless."

Mark Zuckerberg and several Facebook executives went so far as to stage an intervention for Musk, inviting him to dinner at Zuckerberg's house so they could ply him with arguments that AI is okay. It didn't work.

Ever since, Musk and Zuckerberg have waged a social media feud on the topic. Asked about Musk, Zuckerberg told a Facebook Live audience, "I think people who are naysayers and try to drum up these doomsday scenarios—I just, I don't understand it. It's really negative and in some ways I actually think it is pretty irresponsible." Pressed to characterize Musk's position as "hysterical" or "valid," Zuckerberg picked the former.

Musk tweeted in response: "I've talked to Mark about this. His understanding of the subject is limited."

The AI safety debate has become the Pascal's Wager of a secular industry. In the seventeenth century Blaise Pascal decided that he should believe in God, even though he had serious doubts, because the stakes are so high. Why miss out on heaven, or get sent to hell, just to be right about atheism?

In its general form, Pascal's Wager is a classic problem of decision theory. Should a rational person be willing to pay a small cost in order to avoid a chance of a great loss? Prudent people do this all the time when they buy insurance or wear seat belts. But it gets more complicated when the chance of the great loss is hard to evaluate—or when there is dispute over whether the risk exists at all.

To those who work in AI research on a daily basis and have never encountered anything like a control problem, such talk can seem like off-topic scaremongering. To the general public, the risks sound like science fiction or a joke. ("Killer robots" are funny until somebody gets hurt.)

Both sides of the AI debate tout a post-singularity paradise. The difference is over whether there's also an AI hell that we need to watch out for. Should we develop AI and be extra careful (even though many say the caution is unnecessary), or should we develop AI and *not* be so careful (even though many say that would be a terrible mistake)? Who really wants to bet against the possibility of unintended consequences?

The cross talk is fed by differing definitions of "artificial intelligence." There is a distinction between the next generation or two of today's AI and the all-powerful AI that may result from an intelligence explosion. "Would you not invent the telephone because of the possible misuse of the telephone by evil people?" asked Schmidt. "No, you would build the telephone and you would try to find a way to police the misuse of the telephone." Clearly Schmidt is thinking about a different kind of AI, and a different kind of "misuse," than Bostrom and Yudkowsky are.

"The so-called control problem that Elon is worried about isn't something that people should feel is imminent," Gates said. "This is a case where Elon and I disagree." Gates thinks we've got more time to sort out the risks than Musk does.

The more reasonable critics say that no intelligence explosion is on the immediate horizon. Maybe we don't need to get all worked up right now. The nature of the control problem will become clearer with time. We can better deal with it later.

Ever in the background is a concern among tech execs that having the public, and politicians, weigh in on the issue now may result in ham-handed regulation that solves nothing. Poorly crafted regulation could slow progress on a beneficial technology and, perhaps, move research to less scrupulous nations.

Yet no one really has any idea when an intelligence explosion might come. It's not totally inconceivable that a game-changing insight could turn up tomorrow. Even if the original discoverer takes it slow, word would get out. Perhaps a hacker on the other side of the world could write a virus to usurp the processing power of infected computers all over the globe and launch a home-brew intelligence explosion. This may not require institutional support or a Google budget.

The AI-risk think tanks have bright people confronting ethical, political, and philosophical questions that may not be foremost in the minds of AI engineers. The longer the think tanks are able to do that, the more likely they are to come up with conceptual frameworks, options, and solutions that would be useful when an intelligence explosion becomes imminent. Better that than for an isolated team of engineers to have to invent an ethical universe over the weekend that AI becomes all-powerful. As Yudkowsky said, "I don't think we should ignore a problem we'll predictably have to panic about later."

* * *

In 2018 Jeff Bezos and Amazon hosted a conference in Palm Springs where neuroscientist Sam Harris debated MIT roboticist Rodney Brooks (cofounder of iRobot, known for carpet-cleaning robots). Harris expressed concern that competitors in an AI arms race would ignore precautions. "This is something you made up," Brooks objected. Harris had no data; he was saying things that could not be proven either way.

Of course, nobody has data about a technology that doesn't yet exist. Absence of evidence doesn't prove a technology is safe. But Harris was speaking in part about human psychology, something that is not such a total unknown. Should the control problem be solved, super-AI would be the greatest invention of all time, a genie that could potentially grant wishes of wealth, happiness, health, and longevity. This genie could also grant special power, or total power, to whoever got there first. It's hard to see how that wouldn't devolve into a free-for-all among competing academic teams, corporations, and nations.

That was Russian president Vladimir Putin's point in a 2017 talk to a student forum. Whoever is first to master AI will be "ruler of the world," Putin said. He was quick to add that no nation should monopolize AI, and that Russia would share its knowledge with the entire world, just as it does with nuclear technology. It did not take undue cynicism to conclude that Russia had no intention of sharing its AI expertise.

The truly intractable control problem may be the human and political one, of distrustful competitors racing toward the finish line. A winner-take-all mind-set incentivizes rivals to cut corners. If the prudent go slow and the reckless go fast, it's the reckless ones who will launch the first intelligence explosion. And if they get it even slightly wrong, there may not be a bug fix.

You Are Here

On April 29, 2016, a weasel or marten got into the Large Hadron Collider and gnawed through a power cable, shutting it down for several days.

On November 20, 2016, a stone marten hopped over a fence at the LHC and scampered onto an eighteen-thousand-volt transformer, electrocuting itself and causing a short circuit. The collider lost power and shut down.

In June 2006 a "raccoon attack" that "seemed to be a coordinated effort" temporarily shut down the Fermilab.

We live in a world of randomness. Over the past couple of years, I have often invoked that premise to explain the doomsday argument at the dinner table. I've found that I'm always asked two questions: Do I believe these "crazy" predictions? How much time have we got?

My answer to the first question is that I accept Gott's version of doomsday but discount the Carter-Leslie one. I add that this is a moderately unusual position. Most scholarship lumps the two together and focuses on the Carter-Leslie version.

Like all statements of probability, the doomsday arguments are conditional on states of knowledge. Gott works from one big, bold assumption: that we know nothing about the time scale of human existence. I believe this is not an unreasonable assumption, all things considered. The Copernican method requires only cluelessness, an ingredient in abundant supply in our universe.

Even for those who do have strong beliefs about the longevity of our species, Gott's method offers a benchmark. It tells what a rational person without such strong beliefs should think.

Yet Gott's skeptical approach is a hard sell. We think we know stuff. We believe we can handicap an extinction event no one has ever witnessed.

This is presented as the selling point of the Carter-Leslie doomsday argument. It allows me to lay odds on doomsday, and then to adjust those odds for my position in time. The Carter-Leslie scheme is ideally suited to a "Tarzan" who doesn't know his position in time. It tells such a person exactly how much he should shift his beliefs upon learning his birth rank.

But I'm no Tarzan, and neither are you. Most of those aware of the doomsday argument believe we are facing existential risks unique to the present moment. Our ordinary beliefs about the future already incorporate knowledge of our position in time. No further adjustment is necessary.

I think, therefore, that the Carter-Leslie doomsday argument is narrowly correct but not useful in the way that most people expect it to be. I think that Gott's Copernican method contains almost all the wisdom that is to be had from doomsday-style reasoning.

How long have we got? My standard answer is 760 years. That's a median birth-clock Copernican prediction translated into years, using current population estimates. I am surprised at how often the reaction is *relief.* Seven hundred sixty years means it won't affect them, their great-grandchildren, or anybody they know. (I don't bother to mention the year-based median prediction of 200,000 years. Nobody would care.)

I have noticed that hardly anyone contests the 760 years figure as tinfoil-hat crazy. The chart below suggests why. It summarizes some popular and expert estimates of the date of human extinction. There is a lot of overlap.

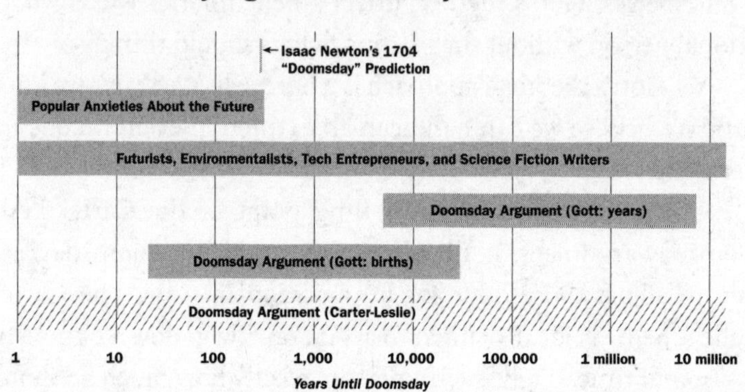

In a 1704 letter, Isaac Newton used the book of Daniel to predict that the end of the world would come in AD 2060. That was then 356 years in the future. Newton offered his forecast in the hope that it would "put a stop to the rash conjectures of fanciful men who are frequently predicting the end of time."

Fat chance of that. But today, when journalists or the public worry about human extinction, they are usually thinking of topical hazards, those affecting the coming decades or centuries—rarely much beyond that.

Futurists, environmentalists, tech entrepreneurs, cosmologists, transhumanists, and science fiction writers offer a gaggle of opinions about when the world will end. Their predictions range from any time now to never. Given the unbounded range of informed opinion, doomsday predictions can hardly be expected to challenge the conventional wisdom. Copernican method predictions say the human race will end in as little as about 20 years (a low-end, birth-clock estimate) or as much as 7.8 million years (high-end, regular-clock). Anyone who strongly disagrees, please raise your hand.

The Carter-Leslie doomsday argument is different. It maps any given set of estimates onto another, more pessimistic set. The Carter-Leslie bar therefore spans the whole range of opinion (indicated by the cross-hatching).

"Personally, I now think we humans will be wiped out this century," Frank Tipler told me. He may be the most pessimistic of Bayesian doomsayers, followed by Willard Wells (who gives the same time frame for the end of civilization and the beginning of the postapocalypse).

But Leslie believes there is a 70 percent chance of long-term survival (beyond five hundred years) after adjusting for the doomsday shift. Bostrom puts the probability of long-term survival at a similar three in four (and does not believe a doomsday shift is justified). Leslie and Bostrom are more optimistic than many who follow the news and pay no heed to Bayes's theorem or killer robots.

It is not so much the human future but rather the fate of intelligent life in the universe where self-sampling may upend the conventional thinking. Enrico Fermi believed in a universe well populated with ETs who are *not* like us, having populations and technological capabilities vastly greater than anything we see on Earth. J. Richard Gott III offers an alternate, more modest assumption, that our situation as humans is not too atypical of observers elsewhere. ETs have not explored and populated vast regions of the universe, and neither have we, because that's not something that many observer-species do.

If typical ETs are not as advanced as we think, then humans are not as primitive. Gott suggests that humans will probably be one of our universe's success stories. We will have survived longer and done more great things than most observer-species out there. It's just that this success may not be what we see in the movies. It is not necessarily our destiny to zip across the galaxy or exist for many millions of years.

Gott's case, made in 1993, has received only moderate traction despite its simplicity and lack of arbitrary assumptions. For much of the scientific and philosophic community, "delta *t*" and "Copernican method" remain fighting words.

Yet past decades have seen a quieter shift in opinion. The space age began with supreme confidence that life and observers are common in the universe. We are only now realizing that we don't know that. We may have been bewitched by a selection effect. Such advocates as Gott, Francis Crick, Brandon Carter, and Nick Bostrom have argued that we cannot exclude the possibility that life, observers, and technological civilizations are vanishingly rare.

The exciting thing is that this is a matter that may be resolved in the coming decades. We are on the threshold of determining whether there is life elsewhere in our solar system. Discovery of life on Mars or Europa or Enceladus would supply a second data point, greatly strengthening the Bayesian case for a universe rich in life. Meanwhile, SETI efforts continue, and the value of negative results should not be discounted. As Gott says, his idea is supported by every SETI project that fails to find ET signals.

There is a striking difference between popular enthusiasm for extraterrestrial life and the trepidation that some of today's scholars feel about it. Should life turn out to exist beyond the Earth *and* SETI never detect signals, it would cast a deep Bayesian shadow on our future. It would raise the probability that great existential risks lie ahead of us. This would be the most important "doomsday argument" of all, one truly telling us something we didn't already know. Bostrom's words reverberate: "dead rocks and lifeless sands would lift my spirits."

No Tree Grows to the Sky

The controversies of self-sampling speak to our present as well as our future. We live in an epoch of exponential growth, in a culture that has convinced itself that such growth will continue a long time. It is our historically growing population that gives the doomsday argument its sting. The ongoing growth of computing power leads to worries about control problems or being simulations in someone else's machine. But all growth spurts must stop sometime. No tree grows to the sky.

So-called doomsday predictions come with generously wide error bars. They can be alarming only to those sold on a particular conception of the future—our galactic manifest destiny. Moore's law, cities on Mars, interstellar probes, and posthuman consciousness are part of many people's mental furniture. You don't have to embrace this cultural infrastructure to be influenced by it. Most of us are too busy with our own lives to think much about remote posterity. It's the ambient culture that shapes prior probabilities. What else can the future be, except what's in the movies?

The doomsday and Fermi controversies are thus stories of cultural expectations out of sync with likely reality. It is not that we are "doomed" or "alone" but that we have been taking the improbable for granted. Gott wrote:

> The odds are against our colonizing the Galaxy and surviving to the far future, not because these things are intrinsically beyond our capabilities, but because living things usually do not live up to their maximum potential. Intelligence is a capability which gives us in principle a vast potential if we could only use it to its maximum capacity, but so does the ability to lay 30 million eggs as the ocean sunfish does. We should know that to succeed the way we would like, we will

have to do something truly remarkable (such as colonizing space), something which most intelligent species do not do.

That could be the most important insight of the global community that has arisen around the doomsday argument, self-sampling, and existential risk assessment. A long human future is not an impossible goal. It may, however, be something that has to be earned by being smarter, wiser, kinder, more careful—and luckier—than we've ever had to be before. The first rule of defying the odds is to never deny the odds.

Early though we may be in the future running through our heads, we are always and already running out of time. Like our remote ancestors, and like all who come after, we see in the distance a singularity, a boundary of the reference class, a monolith marking the end of the world as we know it. We are about to discover the truth of how special we are.

Acknowledgments

Nick Bostrom, J. Richard Gott III, and John Leslie were exceptionally generous with their time, expertise, and patience. James Dreier, Adam N. Elga, and Arnold Zuboff were helpful in tracing the early history of Sleeping Beauty and allied puzzles.

Special thanks go to Tracy Behar, Hendrik Bessembinder, John Brockman, Carleton M. Caves, George Dyson, Peggy Freudenthal, David Goehring, Larry Hussar, Kev L'Baz, Katinka Matson, Arthur Saint-Aubin, Halina and Mike Simm, Ian Straus, Jean and Mark Tansey, Frank J. Tipler, and the staffs of the New York Public Library and the UCLA Research Library. I am further grateful to John and Jill Leslie for their hospitality in putting me up in Victoria, and to Tom Lee for information on *Achilles and the Tortoise.*

Notes

Epigraph

v "Time is a game played beautifully by children": This is Brooks Haxton's 2001 translation of Heraclitus's *Fragments.*

v "He marveled at the fact": Lichtenberg 1990, 113.

Diana and Charles

3 Gott tested prediction method with Charles and Diana: Gott interview, July 31, 2017.

4 Champagne bubbles epitomize chaos theory: Liger-Belair 2004.

How to Predict Everything

9 nuclear accidents, use of Bayes at RAND: McGrayne 2011, 124–28; Klepper 2003.

10–11 "more dramatic incidents... a probability that is very small": Iklé, Aronson, Madansky 1958.

12 Fallout would have reached Philadelphia: Mosher 2017.

12 "Until my death": Tuttle 2013.

13 Bayes adopted by insurance companies: McGrayne 2011, 42–45.

14 *"I met a traveller from an antique land":* bit.ly/2KyfkQO.

15 Berlin Wall prediction: Gott interview, July 31, 2017.

16 "In the age of Quantum Mechanics": Sowers 2002, 44.

20 79 percent of Broadway musicals: Bresiger 2015.
21 "How to Predict Everything": Ferris 1999.
22 "According to a law established": Goldman 1964, 34.

Riddle of the Sphinx

24 200,000 years ago: This is a widely adopted figure, used in Gott 1993. In 2017 there were reports of fossil skulls from Morocco, dated 300,000 years ago. Should this claim hold up, it would increase the Copernican estimate by 50 percent. The dating of the Morocco find has been questioned.
25 5,100 to 7.8 million years: Ferris 1999; Gott interview, July 31, 2017.
25 1 to 2 million years lifespan for mammal species: Lawton and McCredie 1995.
26 "Or consider lemmings": Leslie 2010, 459.
27 70 billion: In his 1993 *Nature* article Gott used 70 billion. Over 3 billion have been born since then. Meanwhile, estimates of the cumulative population have crept upward. In 2011 the Population Reference Bureau estimated the cumulative population at 108 billion.
27 12 to 18,000 years: There is a typo in Gott 1993, giving the range as 12 to 7.8 million years. The upper figure should be 18,000 with the population figures Gott used.
28 "I am loath to tell people": Brantley 2015.
29 The full Carter-Leslie version: Leslie 1996.
30 *99 percent:* The probability of doom soon, given an early birth rank, is $p/(p+(1-p)\times(soon/late))$. Here p is the prior probability of doom soon, and *soon/late* is the population ratio of doom soon to doom late.

The Minister of Tunbridge Wells

32 "I never saw a worse collection": Bellhouse 2004, 12.
32 "the variety of persons and characters": Bellhouse 2004, 12.
33 Bayes family and cutlery business, Sheffield: Bellhouse 2004, 3.
33 Stanhope nominated Bayes to Royal Society: Bellhouse 2004, 13.
34 "No testimony is sufficient": Hume 1748.
34 Theorized that Bayes was motivated by Hume: Stigler 2013.
35 "The purpose I mean is": Bayes 1763.
36 only source: The crucifixion rates passing mention in the *Annals* of Roman historian Tacitus. He records that "Christus," namesake of "a

class hated for their abominations, called Christians...suffered the extreme penalty during the reign of Tiberius at the hands of one of our procurators, Pontius Pilatus."

36 Price's loophole: See Stigler 2013.

37 "When you hear hoofbeats": The quote has been attributed to various authors and has been identified as "an old Yiddish saying." See bit.ly/2lsKMVw.

37 "Don't assume an observation": Leslie interview, January 17, 2018; see also Leslie 2010, 447.

38 Urn example: See Bostrom 2002, 97.

40 "Rational belief is constrained": Bostrom 2002, 78.

40 Laplace as true originator: McGrayne 2011, 22–33.

42 270, 276 tanks a month: See Wikipedia entry for the "German tank problem," bit.ly/25O0xXE.

A History of Grim Reckoning

43 "So I picked up the *New York Times*": Gott interview, July 31, 2017.

44 "Copernican Cosmological Principle" in 1952 book: Bondi 1952.

44 "Copernicus taught us the very sound lesson": Carter 1974.

45 Eddington's fishnet example: Eddington 1939, 16–37.

45 "Whenever one wishes to draw general conclusions": Carter 2004, 2.

46 "Anthropic notions flourish in the compost": Brown 1988, a review of John D. Barrow and Frank J. Tipler's *The Anthropic Cosmological Principle.*

46 Physicists hissed: See Tegmark 2014, 144, which cites an incident at the Fermilab in 1998.

46 "not something that I would be prepared": Carter 1983, 141.

46 "Intelligent information-processing": Barrow and Tipler 1986, 23.

46 Completely Ridiculous Anthropic Principle: Gardner 1986.

46 "an application of the anthropic principle outstandingly free": Leslie 1996, 193.

47 Carter did not accept grim forecast: Frank J. Tipler personal email, April 5, 2018.

47 "world's greatest expert": See video of Holt's conversation with Leslie, bit.ly/2FLfyAq; Leslie and Kuhn 2013.

48 1987: Email, John Leslie to Toby Ord, April 10, 2017, supplied by Leslie.

48 "Tipler was one of my special pals there": Leslie interview, January 17, 2018.

48 Omega Point Theory class: See Tipler's Tulane web page, bit.ly/2IKlteG.

48 "became convinced of its importance": Leslie personal email, October 24, 2017.

49 "Carter-Leslie doomsday argument...not only the credit": Leslie 1996, 188.

49 "muttering support from the trenches": Leslie interview, January 17, 2018.

49 using the word "doomsday": I have been unable to discover who coined the term "doomsday argument." Leslie was under the impression that Tipler invented it, but Tipler told me he didn't, and he doesn't know who did (Tipler personal email, April 6, 2018). As far as I can tell, Nielsen's 1989 article is the first to use "doomsday" in this connection. Though some regard the name as regrettably grim, it is now entrenched.

49 "Now my point is that this procedure": Nielsen 1989, 456.

49 "It is a pleasure to thank N. Brene": Nielsen 1989, 467.

50 Saw Berlin Wall demolition on TV, called friend: Gott interview, July 31, 2017.

50 "Chuck, you remember that prediction": Tyson, Strauss, Gott 2016, 413; Gott interview, July 31, 2017.

50 "I thought, well, you know": Ferris 1999.

51 "the location of your birth": Gott 1993, 316.

51 *self-sampling assumption; human randomness assumption:* The first is Nick Bostrom's term, the second William Eckhardt's. See Bostrom 2002 and Eckhardt 1997.

51 "Disturbingly, even extraordinarily low values": Gott 1993, 317.

51 "The methods that I have used here": Gott 1993, 319.

51 1 in a billion: Gott 1993, 318, and personal email, June 27, 2018. Specifically, Gott said that the chance humans will settle on a billion planets is 1 in a billion "because the chance you would randomly find yourself in the first one-billionth of all habitation sites ever occupied by humanity is 1 in a billion."

52 "'There are lies, damn lies and statistics'": Goodman 1994, 106.

52 Feature in *New York Times:* Browne 1993.

52 "pseudo-science, a mere manipulation of numbers": Lerner 1993.

52 "Mr. Lerner refuses": Gott 1993a.

53 "After careful consideration": Dyson 1996.

53 Leslie defended his book in a letter to *Nature:* Leslie 1997.

53–54 "apparently under the influence"; "I have found however that such conclusions": Carter 2006, 5.

54 Technologically adept... universe cooled off to absolute zero: Several recent physical findings seem to rule out the literally infinite feature of Dyson's scheme. The universe may not cool off to absolute zero. For an accessible description see J. Richard Gott's in Tyson, Strauss, Gott 2016, 406–407.

54 "a universe growing without limit": Dyson 1979, 459–460.

54 "The anthropic principle's attribution of comparable a priori weighting": Carter 2006, 5.

55 future will be postapocalyptic: Wells develops this thesis in Wells 2009.

55 860 billion person-years: Wells 2009, 85.

56 "And so the short answer": Wells 2009, 120.

Twelve Reasons Why the Doomsday Argument Is Wrong

57 "Is the Doomsday Argument correct?": Dieks 1992, 79.

57 "I have encountered over a hundred objections": Bostrom 2002, 125.

57 "Given twenty seconds" and "At least a dozen times": Leslie 1996, 206 and 219.

58 "big question... whether we have any right": Leslie 1996, 203.

59 Self-sampling and traffic: Bostrom 2002, 82–84.

60 Adam and Eve: Almost everyone who thinks about the doomsday argument ends up asking a version of this question. Nick Bostrom used the Cro-Magnon example (Bostrom 2002, 116). John Leslie considered an ancient Roman (Leslie interview, January 17, 2018; Leslie 1996, 205.)

61 Emerald experiment: Leslie 1996, 20.

65 "Like any good scientific hypothesis": Gott 1993, 319.

66 "Dear Mr. Newton: What's all this nonsense": Leslie 1996, 219.

69 "In the absence of data": Sober 2003, 420–421.

69 "you can't get any result from a single trial": Leslie interview, January 17, 2018.

69 "Will you repeat the experiment": Leslie interview, January 17, 2018.

Twenty-Four Dogs in Albuquerque

71 "incredibly irresponsible"; "Anybody can see it's garbage": Caves interview, December 12, 2017.

71 "Gott dismisses the entire process": Caves 2000, 2.

71 "it was important to find": Caves 2000, 2.
71 "a notarized list of…24 dogs": Caves 2000, 15.
72 "Gott is on record as applying": Caves 2008, 2.
73 "We can distinguish two forms": Bostrom 2002, 89.
73 "I didn't put any Bayesian statistics": Gott interview, July 31, 2017.
75 "When you can't identify any time scales": Caves 2008, 11.
77 "No other formula in the alchemy of logic": Keynes 1921, 89.
77 Goodman's objection to Gott: Goodman 1994.
79 Jeffreys prior compatible with location- and scale-invariance: This fact was demonstrated not by Jeffreys but by Washington University physicist E. T. Jaynes. See Jaynes 1968.
79 Jeffreys prior: The usual way of describing the Jeffreys prior is to say that the probability of any numerical value N is proportional to $1/N$.
79 "For each of the six dogs": Caves 2000, 15.
79 "I just don't do bets": Glanz 2000.
79 "It is inescapable that he doesn't believe": Glanz 2000.
80 art collector; "enough to buy a very nice piece": Caves interview, December 12, 2017; Caves 2008, 11.
80 "the intervals that Gott finds": Caves 2000, 15.
80 Joke about green bananas: Gott interview, July 31, 2017.
80 "I went to the *Guinness Book of World Records*": Gott interview, July 31, 2017.
81 $1,100 in wallet: Coughlan 2012. The film was *World War Z* (2013).

Baby Names and Bomb Fragments

82 thousand-year Reich: Gott interview, July 31, 2017
82 White Sox prediction: Glanz 2000; Caves 2008, 2.
83 1981 monuments calendar: Gott interview, July 31, 2017.
84 *The Bohemian Girl:* Some theatrical companies put on short productions to fill gaps in a theater's schedule. *The Bohemian Girl* is an example, having opened just two nights earlier and closing with the January 9 performance. The future run of *The Bohemian Girl* was therefore zero, and zero can't be charted on a logarithmic scale. It counts as a wrong prediction for the Copernican method, which would have predicted at least 2/39 more nights.
86 "looks unlikely to ever stop": Saunders 2015.
87 Examples of Zipf's law: See Wells 2009, 52.

89 Zipf's law is closely related to Copernican method/Lindy's law: The point is made in Wells 2009.
91 Value of stock as future income stream: Williams 1938.
92 Ben Reynolds's example of stock valuation: Reynolds 2016.
93 Berkshire Hathaway returns: See 2014 shareholder letter, bit.ly/1g0xVxs.
94 Berkshire Hathaway stock holdings: See bit.ly/2KBQ0tb. Ages of companies are from Wikipedia, taking the older of the predecessor firms in cases of recent mergers.
94 "be prospering a century from now": 2014 shareholder letter, bit.ly/1g0xVxs.
94 "Time is the friend of the wonderful company": Reynolds n.d.
94 "At Berkshire, we make no attempt": Warren Buffett, quoted in Connors 2010, 157.
95 most stocks have done *worse* than US Treasury bills: Bessembinder (forthcoming).
95 Life of exchange-traded stock barely seven years: Bessembinder (forthcoming).
97 "If there's something in the culture": Taleb 2012.
98 Harry Potter books in San Diego library: Wells 2009, 14.
98 Comparison between runs of plays and companies: Wells 2009, 3.

Sleeping Beauty

102 "are structurally the same": Dieks 2007, 13.
103 as early as 1983; early history of "Sleeping Beauty": Arnold Zuboff personal email, April 21, 2018; Zuboff 1990; Adam Elga personal email, April 20, 2018; James Dreier personal email, April 18, 2018. See also Dreier's post at bit.ly/2rmtagm; archived rec.puzzles posts at bit.ly/2IbURmG; Elga 2000; Zuboff 1990. Sleeping Beauty is closely related to the "paradox of the absentminded driver," a problem described by economists Michele Piccione and Ariel Rubinstein in a 1997 paper. (The motorist has to choose the right freeway exit to get home and can't remember how many he's passed.) See Piccione and Rubinstein 1997, especially 12–14.
103 "a game . . . being played in an amazingly big hotel": Zuboff 1990, 20.
105 "probability is the wrong tool": Armstrong 2017, 4.
106 Intuition fails on amnesia pills: See Armstrong 2017, 4.
106 "There are tricky problems": Armstrong 2017, 4.
107 live for today: Armstrong 2017, 8. Armstrong calls this being "psychologically selfish." He distinguishes it from being "physically selfish." A

physically selfish person cares about all observer-moments attached to his physical body. Thus he would cooperate with earlier or later awakenings.

107 "Sailor's Child": Neal 2006, 17–18.

The Presumptuous Philosopher

111 "If you don't sign up your kids for cryonics": Yudkowsky's LessWrong post, "Normal Cryonics," bit.ly/2KN5gUk.

111 "His interest in science was a natural outgrowing": Khatchadourian 2015.

111 "the ever accelerating progress of technology": Ulam 1958.

112 "seemed interesting and important": Bostrom interview, November 17, 2017.

112 "thought it was a crazy topic": Bostrom interview, November 17, 2017.

112 "One should reason as if": Bostrom 2002, 57.

113 "I, the elephant, wrote this": Book 8 of Pliny's *Natural History.*

115 Uploads would make doomsday prediction ambiguous: William Eckhardt makes this point (with "human brains inside of robots") in Eckhardt 1993.

115 "When I started thinking about what's really wrong": Bostrom interview, November 17, 2017. Like the doomsday argument, SIA had multiple independent originators. Tomás Kopf, Pavel Krtous, and Don M. Page published a version of SIA in 1994, and Paul Bartha and Christopher Hitchcock did so in 1999.

115 "Given the fact that you exist": Bostrom 2002, 66.

116 wakenings count as observers: Bostrom's strong self-sampling assumption (SSSA) uses observer-moments rather than observers. It is more flexible and generally to be preferred. In Sleeping Beauty an hour-long wakening could count as an observer-moment.

117 "As an application of Bayesian reasoning": Dieks 2001, 16.

118 "Those Nigerian email solicitations": See bit.ly/2sr81Ci.

119 "I thought some more and thought": Bostrom interview, November 17, 2017.

120 "Sometimes it is Bostrom who is presumptuous": Gerig 2012, 7.

120 Olum defends self-indication assumption: Olum 2000.

120 "Entities must not be multiplied beyond necessity": This phrasing (translated from Latin) is credited to Irish philosopher John Punch. It dates from 1639, three centuries after English theologian William of Ockham (c. 1287–1347).

Tarzan Meets Jane

121 "The theory of probability is, in the most profound way": Translation by Arthur Saint-Aubin. The passage is cited in Gorroochurn 2011, which provides a rogues' gallery of learned error in probability theory.

121 D'Alembert and probability errors: Gorroochurn 2011. Two centuries later statistician Karl Pearson wrote: "What then did D'Alembert contribute to our subject? I think the answer to that question must be that he contributed absolutely *nothing*."

122 "Reasoning About the Future: Doom and Beauty": Dieks 2007.

122 "If a certain piece of evidence": Dieks 1992, 80.

124 "a bizarre situation": Dieks 2007, 5.

124 a thousand times more likely to find yourself in the tails world: See Dieks 2007, 432, for full derivation. Dieks uses algebra, not SIA, to arrive at this conclusion.

The Shooting Room

128 "I spent two and a half years": Leslie interview, January 17, 2018.

128 "because I was so near to ending up": Leslie interview, January 17, 2018.

128 "There was a group of his grad students": Leslie interview, January 17, 2018.

129 "a good, hard paradox": Bostrom 1998. See also Leslie 1996.

129 "*Abandon all hope, you who enter this room!*": Bartha and Hitchcock 1999, 404.

130–31 George and mother, Tracy: Bartha and Hitchcock 1999a, 404–405.

132 Improbable versus dangerous; tiger example: Zuboff 2008, 13.

134 "The issue of determinism": Eckhardt 1993.

135 "As long as the validity": Eckhardt 1993.

The Metaphysics of Gumball Machines

136 "Good systems tend to violate": Selby 2013.

136 Dennis turned $5,000 into $100 million: Carr 2018.

137 Eckhardt, Dennis, and the Turtles: Faith 2003.

137 $175 million in five years: This is reported by a former Turtle, Russell Sands. See Carr 2018.

138 No probability paradoxes, only fallacies: Eckhardt 2013.

138 Boss tells you to count balls in urn: Sowers 2002, 41.
138 Eckhardt's numbered-token dispenser: Eckhardt 1993 and 1997; see also Franceschi's elaboration in Franceschi 2009.
139 2009 article: Franceschi 2009.
141 "A Shooting-Room View of Doomsday": Eckhardt 1997.
141 "Betting Crowd" story: Eckhardt 2013.
142–43 "Ninety percent of all players": Eckhardt 2013.
143 "air of paradox": Eckhardt 2013.
143 Chance of civilization surviving twenty-first century is 50 percent: Rees 2003.
144 "There may exist a plethora": Eckhardt 1993, 7.
144 "The Doomsday argument does not fail": Bostrom 2002, 204.
144–45 "Paradoxical applications are distinguished...I wish to suggest that insensitivity": Bostrom 2002, 185 and 202.
145 "quite weak": Bostrom 2002, 202.
145 "The last stretch of the PhD thesis": Bostrom interview, November 17, 2017.
145 "I feel that the problem of the reference class": Bostrom 2002, 205.

The Simulation Hypothesis

150 "accurate-ish": Zullo 2016.
150 "We would be drooling, blithering idiots": Moskowitz 2016.
150 "If I were a character in a computer game": Moskowitz 2016.
150 "effectively zero": Moskowitz 2016.
151 "The strongest argument for us being in a simulation": Bilton 2016. This quote has appeared in many places with slight variations. In some accounts Musk puts the chance of being in "base reality" at 1 in billions.
151 "two tech billionaires have gone so far": Friend 2016.
151 Speculation that Elon Musk was one of the billionaires: Kriss 2016.
151 "the tech industry is moving into territory": Kriss 2016.
151 "basically a religious belief system": Bilton 2016.
151 Gosse biography; invented aquarium; corresponded with Darwin: Thwaite 2002. Gosse coined the word *aquarium*. He was not just interested in fish; he had a particular passion for sea anemones, and his writings sparked a mania for saltwater aquaria. Gosse came to regret that so many sea anemones spent cramped lives in Victorian parlors.
155 10^{33} to 10^{36} operations per second: Bostrom 2003, 248.

156 "a [planet-mass] computer could simulate": Bostrom 2003, 249.
156 Edith Cumbo biography: See the Colonial Williamsburg website, bit.ly/2IxYyQo.
157 *Bodleian Plate:* bit.ly/2rl6Pkl.
158 480,000 visitors a year: Zullo 2016.
159 "a mind in exactly the same sense": Searle 1999.
159 If simulated people have real feelings: Stanislaw Lem's 1974 story, "The Seventh Sally, or How Trurl's Own Perfection Led to No Good," explores this theme. It tells of a well-meaning robot who builds a miniature city for the sadistic amusement of an evil dictator. The robot, Trurl, reasons that the dictator can oppress the miniature city rather than a real city. But Trurl has made a dreadful mistake. The simulation is too real; the little people have real feelings, and the dictator's simulated oppression is just as bad as the real kind.
160 "Roko's Basilisk": Auerbach 2014. A Basilisk is a medieval monster. Look at it and you die. The contemporary version, a monster from the future, is named for a poster with the screen name "Roko" on the LessWrong site founded by Eliezer Yudkowsky. Yudkowsky deleted the post as an "information hazard," inadvertently ensuring its memehood. The post is archived on RationalWiki (bit.ly/2rvQqsv). The dreadful Basilisk is credited for one celebrity romance. Elon Musk met Canadian musician Grimes because of a shared interest in the concept.
161 "quite weak": Bostrom 2002, 202.
164 In the absence of data: Sober 2003, 420–421.
164 Silas Beane and colleagues: Beane, Davoudi, Savage 2012.
164–65 "Should any error occur": Bostrom 2003, 5.
165 "Your motivation to save for retirement": Hanson 2001.

The Fermi Question

166 "Where is everybody?": Poundstone 1999, 22.
167 "In the middle of this conversation": Jones 1985.
167 "What we all fervently hope": Fermi 2004.
167 "absolutely certain (1) that there would be a nuclear war": Putnam 1979, 114.
168 Marconi believed he picked up signals from Mars: See "Marconi Sure Mars Flashed Messages," *New York Times,* September 2, 1921.
168 Frank Drake's 1960 attempt: Poundstone 1999, 50–51.

168–69 Drake equation: Poundstone 1999, 54–59.
169 More than 3,800 planets: The NASA Exoplanet Archive keeps a running tally: bit.ly/1H3HJVw.
170 "Absence of evidence": Oliver and Billingham 1971, 3. Internet quote mills now attribute the aphorism to a wide range of thinkers, from Carl Sagan to Donald Rumsfeld. Sagan (who credited Rees) invoked it repeatedly.
170 "belongs back where it came from": Purcell 1963. The statement was first made in a 1960 speech at Brookhaven National Laboratory.
170 John Ball's "zoo hypothesis": Ball 1973.
172 a million years for von Neumann probes: Brin 1983, 283–284.
172 Fermi paradox is question mark over single data point: Bostrom 2002, 16.

The Princess in the Tower

173 "This, to me, speaks rather persuasively": Poundstone 1999, 145.
174 "In order to display the difference": Poundstone 1999, 145.
174 "I do not know whether such a line of reasoning": Francis Crick, quoted in Carter 1983, 139.
180 "there are many paths": Poundstone 1999, 145.
181 Evolution of wings: Carter 1983, 359.
181 Carter's mathematical analysis, one or two crucial steps: Carter 1983.
181 Hanson's computer simulations, five crucial steps: Hanson 1998.

Two Questions for an Extraterrestrial

188 "If you believe that our intelligent descendants": Gott 1993, 319.
191 more than forty orders of magnitude: Sandberg, Drexler, Ord 2018, 5–6.
192 "probably extremely far away": Sandberg, Drexler, Ord 2018, 16.
192 Clarke's words: "Any sufficiently advanced technology is indistinguishable from magic" is presented as Clarke's Third Law in *Profiles of the Future* (1973). (Hence a self-reproducing robot can look like a monolith.)
193 Einstein was a smart guy: Gott interview, July 31, 2017.
193 "beyond which human affairs": Ulam 1958.

Pandora's Box

194 consensus talking points shifted: John Leslie made this point. See Leslie 2010, 457.

195 CERN physicist arrested for ties to Al-Qaeda: Clark and Overbye 2009.

195 Baguette dropped by bird: Page 2009.

195 GOD IS SABOTAGING: *University Post* (at University of Copenhagen), October 19, 2009.

195 TIME-TRAVELING HIGGS SABOTAGES THE LHC: *New Scientist,* October 13, 2009.

196 "a pair of otherwise distinguished physicists": Overbye 2009.

196 American collider plans dropped: Nielsen and Ninomiya 2009.

196 "Our theory suggests": Overbye 2009. Nielsen and Ninomiya's idea is similar to that in a 1985 science fiction story, "The Doomsday Device," by physicist John Gribbin. Gribbin's tale also involves a supercollider that runs into a series of accidents preventing its operation.

196 "It is not far-fetched to suppose": Khatchadourian 2015.

198 "It would be great news to find that Mars": Khatchadourian 2015. See also Bostrom 2002, 16.

198 Collects news clippings of catastrophes: Leslie interview, January 17, 2018.

198 Reports of a contraceptive: Leslie 2010, 452–453.

199 "I want to live in a world": Lee 2017.

200 "ice-nine": As described in Kurt Vonnegut's 1963 novel *Cat's Cradle,* ice-nine is a type of ice that is solid at room temperature. A speck of it transforms any body of water into ice-nine.

200 "the ultimate ecological catastrophe": Coleman and De Luccia 1980, 3314.

201 "structures capable of knowing joy": Coleman and De Luccia 1980, 3314.

203 "There is something so clean": Eckhardt 1993, 7.

203 "good-story bias": Bostrom 2002a, 18.

204 "*I Love Lucy* has spread": Brin 1983, 297.

204 2005 article on cosmic-scale catastrophes: Tegmark and Bostrom 2005 and 2005a.

204 "that precludes any observer"; "a false sense of security": Tegmark and Bostrom 2005, 1.

205 six inches across: The externally measured diameter of a black hole is directly proportional to its mass. If the Earth were a black hole, it would be barely one-third of an inch across. Neptune is about seventeen times more massive than the Earth.

207 Arkani-Hamed study: Arkani-Hamed, Dubovsky, Senatore, Villadoro 2008.

208 "Our basic result": Tegmark and Bostrom 2005a, 3.

Life and Death in Many Worlds

209 “I think Einstein should stop”: O’Connell 2014.

209 LHC card game: Nielsen and Ninomiya 2009.

210 $4.75 billion to build LHC: Knapp 2012. The yearly operating budget is now about $1 billion.

210 “seem lunatic”: Deutsch 2011, 310.

212 A sizable fraction believe many worlds: See Tegmark 1997, 1. Other polls—inevitably described as “nonscientific” despite the fact that they’re polling scientists—have claimed as much as 50 percent support.

213 “shut up and calculate”: Max Tegmark credits the phrase to Anupam Garg. See Tegmark 1997, 4.

213 “like a giant smooth mountain”: Merali 2015.

214 *Get in the box with Schrödinger’s cat:* Papineau 2003.

215 “I’d be OK”: Chown 1997, 51.

215 “Your cranial explosive will be intact”: Moravec 1988, 190.

215 Everett’s views on quantum immortality: Shikhovtsev 2003, 21.

216 “When one fateful day”: Tegmark 2014, 220.

218 “a postmodern fanatical religious cult”: Mallah 2009.

218 “people have considered putting the experiment”: Mallah 2009.

218 going to a parallel universe to be with her father: Shikhovtsev 2003, 21. This was reported by Glenn Fishbine.

218 “Don’t try this at home”: Tegmark 2014, 220.

219 sixty-eight clicks; meteorite scenario: Tegmark 2014, 219.

219 “dying isn’t a binary thing”: Tegmark 2014, 219.

222 “Many physicists would undoubtedly rejoice”: Tegmark 1997, 5.

223 “total amount of consciousness”: Mallah n.d.

1/137

224 “one of the greatest damn mysteries”: Feynman 1985.

224 Eddington and the fine-structure constant: Kragh 2003 has an amusing account.

225 two-dimensional world: Victorian schoolmaster Edwin Abbott Abbott’s short fantasy novel *Flatland: A Romance of Many Dimensions* (1884) initiated thinking about 2-D life. It has been followed by many more earnestly scholarly treatments.

226 Whitrow's argument for the necessity of 3-D space: Whitrow 1955, 31.
226 Cosmological constant is 10^{120} times bigger: Olum and Schwartz-Perlov 2007.
226 "a magic number"; "hand of God": Feynman 1985, 129.
227 "You must not fool yourself": Feynman as told to Leighton 1985, 343.
227 Selection process is different from bearded patriarch: John Leslie has argued that we ought to take seriously the possibility of a Platonic principle (not necessarily a "god" in the traditional sense) selecting universes compatible with observers. See Leslie 1989a.
228 Thomas Digges and Giordano Bruno: An infinite universe was described in Digges's *A Prognostication everlasting,* 1576, and Bruno's *De l'Infinito, Universo e Mondi,* 1584.
229 Bubble universes proposed: Gott 1982; Linde 1982; Albrecht and Steinhardt 1982.
230 Everything would be repeated: Knobe, Olum, Vilenkin 2006.
231 Large, though finite, limit on human variation: Neal 2006.
231 10 to the power of 30 billion: Neal 2006, 24.
231 another dog wager: J. Richard Gott recounts this in Tyson, Strauss, Gott 2016, 398.
232 Hawking's inflation model: Hawking and Hertog 2018.
233 Is the multiverse real?: Bostrom 2002, 11–41, provides an overview.
234 "inverse gambler's fallacy": Hacking 1987.
234 D'Alembert believed gambler's fallacy: Gorroochurn 2011.
234 Psychological data suggests that we're all prone to such beliefs: This is the theme of my 2014 book, *Rock Breaks Scissors* (Poundstone 2014).
235 Leslie one of first to point out error: Leslie 1988. The editor of *Mind* told Leslie that over two hundred scholars submitted refutations of Hacking's article, a record for the journal (Leslie personal email, June 28, 2018).
235 A more proper analogy: See similar examples in Leslie 1988, 270; Bradley 2007, 139.
236 "Life is finite": Wigner and Szanton 1992.
236 "the bigger sleep of never existing at all": Zuboff 2008.
236 "not picky" and "picky": I draw from Bradley 2007, 141, who uses the word "picky" in this context. ("We weren't picky…we are maximally picky.") Arnold Zuboff refers to "subjective" and "objective individuation."
238 breaking down the evidence into two parts: Bostrom 2002, 35.

239 effects cancel out: See Bostrom 2002, 34.
239–40 Imagine that we are alone in the universe: Scharf 2014, 33–35.

Summoning the Demon

242 Arkhipov and Petrov prevented World War III: *Military History Now* 2013.
242 Musk donated $10 million: Dowd 2017.
242 "We cannot necessarily rely": Bostrom 2002.
243 "Let an ultraintelligent machine be defined": Good 1965.
244 Good biography; consulted on *2001:* van der Vat 2009. Nick Bostrom, the former artist, designed the Future of Humanity Institute's logo as an homage to the Kubrick film. The logo is a black, slightly convex diamond that Bostrom likens to the *2001* monolith rotated forty-five degrees.
244 Good said "survival" should be replaced with "extinction": Barrat 2013.
245 "With artificial intelligence, we are summoning": Dowd 2017.
246 "OK, let's get back to work summoning": Dowd 2017.
246 "1. A robot may not injure": The three laws first appear in Asimov's 1942 short story "Runaround" (where they are attributed to the "*Handbook of Robotics,* 56th edition, 2058 AD").
247 "paperclips of doom": See Bostrom 2014, 107–108. Elon Musk tells a similar tale: "Let's say you create a self-improving A.I. to pick strawberries, and it gets better and better at picking strawberries and picks more and more and it is self-improving, so all it really wants to do is pick strawberries. So then it would have all the world be strawberry fields. Strawberry fields forever." See Dowd 2017.
248 "mistakenly elevate a subgoal": Bostrom 2002.
249 "How…do you encode the goal functions": Dowd 2017.
249 "I'm not sure": Dowd 2017.
251 "Robots are invented": Dowd 2017.
252 "This is not an honest conversation": Khatchadourian 2015.
252 "I am in the camp that is concerned": Reddit "Ask Me Anything" session, bit.ly/2jv3DhF.
252 "Frankenstein complex": Khatchadourian 2015.
252 "I don't buy into the killer robot [theory]": Hunter 2017.
252 "I am in the camp that it is hopeless": Khatchadourian 2015.
252 Zuckerberg intervention for Musk: Metz 2018.

252 "I think people who are naysayers": Clifford 2017.
252 "hysterical" or "valid": Dowd 2017.
253 "I've talked to Mark": Clifford 2017.
253 Pascal's Wager: Both Bostrom and Yudkowsky have explored this analogy in the context of how AI should act under uncertainty. When gains are infinitely great, then even infinitesimally small risks should matter to a rational decision maker. Yet this can lead to seemingly absurd cases, as in the paradox of "Pascal's mugging" (see Bostrom 2009). However, the risks of an intelligence explosion seem to be a good deal greater than infinitesimal, and it is not necessary to assign infinite values to a post-human heaven or hell in order to conclude that caution is in order.
253 "Would you not invent the telephone": Ha 2018.
254 "The so-called control problem": Stevenson 2017.
254 "I don't think we should ignore a problem": Horgan 2016.
255 Jeff Bezos and Palm Springs meeting: Metz 2018.
255 "This is something you made up": Metz 2018.
255 "ruler of the world": Fingas 2017.

You Are Here

256 Weasel gnawed cable: Brumfiel 2016. Engineers found "the charred remains of a furry creature near a gnawed-through power cable." Brumfiel wrote that "it is unclear whether the animals are trying to stop humanity from unlocking the secrets of the universe."
256 November stone marten incident: Hersher 2017.
256 "raccoon attack": Lee 2006.
258 "put a stop to the rash conjectures": *Evening Standard* 2007.
258–59 20 to 7.8 million years: Gott 1993. The chart's bar labeled "Gott 1993: births" shows 95 percent limits updated to current population figures.
259 "Personally, I now think we humans": Tipler personal email, April 6, 2018.
259 70 percent chance of long-term survival: Leslie holds that "the world is probably indeterministic in ways that considerably weaken any Bayesian shift towards thinking that doomsday is imminent" (personal email, June 28, 2018).
259 Bostrom puts probability at three in four: Bostrom 1996, 5.
260 second data point: Spiegel and Turner 2012.
260 "dead rocks and lifeless sands": Khatchadourian 2015.
261 "The odds are against our colonizing the Galaxy": Gott 1993, 319.

Sources

Aaronson, Scott. "Can Quantum Computing Reveal the True Meaning of Quantum Mechanics?" 2015. to.pbs.org/269vk1k.

Albrecht, Andreas, and Paul J. Steinhardt. "Cosmology for Grand Unified Theories with Radiatively Induced Symmetry Breaking." *Physical Review Letters* 48 (1982): 1220.

Arkani-Hamed, Nima, and Sergei Dubovsky, Leonardo Senatore, and Giovanni Villadoro. "(No) Eternal Inflation and Precision Higgs Physics." *Journal of High Energy Physics* 03, no. 075 (2008).

Armstrong, Stuart. "Anthropic Decision Theory for Self-Locating Beliefs." Future of Humanity Institute. September 2017. bit.ly/2L2dhnH.

Auerbach, David. "The Most Terrifying Thought Experiment of All Time." *Slate*, July 17, 2014.

Bacon, Francis. *Novum Organum.* 1620. bit.ly/2jojej4.

Bains, William, and Dirck Schulze-Makuch. "The Cosmic Zoo: The (Near) Inevitability of the Evolution of Complex, Macroscopic Life." *Life* 6, no. 25 (2016). doi:10.3390/life6030025.

Baldwin, John, Lin Bian, Richard Dupuy, and Guy Gellatly. "Failure Rates for New Canadian Firms: New Perspectives on Entry and Exit." Ottawa: Statistics Canada (2000).

Ball, John A. "The Zoo Hypothesis." *Icarus* 19 (1973): 347–349.

Barrat, James. *Our Final Invention: Artificial Intelligence and the End of the Human Era*. New York: St. Martin's, 2013.

Barrow, John D., and Frank J. Tipler. *The Anthropic Cosmological Principle.* New York: Oxford University Press, 1986.

Bartha, Paul, and Christopher Hitchcock. "No One Knows the Date or the Hour: An Unorthodox Application of Rev. Bayes's Theorem." *Philosophy of Science (Proceedings)* 66 (1999): S329–S353.

———. "The Shooting-Room Paradox and Conditionalizing on Measurably Challenged Sets." *Synthese* 118 (1999a): 403–437.

Bayes, Thomas. "An Essay Towards Solving a Problem in the Doctrine of Chances." *Philosophical Transactions of the Royal Society of London* 53 (1763): 370–418.

Beane, Silas R., Zohreh Davoudi, and Martin J. Savage. "Constraints on the Universe as a Numerical Simulation." 2012. arXiv:1210.1847.

Bellhouse, D. R. "The Reverend Thomas Bayes FRS: A Biography to Celebrate the Tercentenary of His Birth." *Statistical Science* 19 (2004): 3–32.

Bessembinder, Hendrik. "Do Stocks Outperform Treasury Bills?" *Journal of Financial Economics,* forthcoming.

Biello, David. "The Origin of Oxygen in Earth's Atmosphere." *Scientific American,* August 19, 2009.

Bilton, Nick. "Silicon Valley Questions the Meaning of Life." *Vanity Fair,* October 13, 2016.

Bondi, Hermann. *Cosmology.* Cambridge: Cambridge University Press, 1952.

Bostrom, Nick. *Anthropic Bias: Observation Effects in Science and Philosophy.* New York: Routledge, 2002.

———. "Are You Living in a Computer Simulation?" *Philosophical Quarterly* 53 (2003): 243–255.

———. "Beyond the Doomsday Argument: Reply to Sowers and Further Remarks." n.d. bit.ly/2js28AY.

———. "The Doomsday Argument: A Literature Review." 1998. bit.ly/2IULsNp.

———. "Existential Risks." *Journal of Evolution and Technology* 9 (2002a): 1–31.

———. "Investigations into the Doomsday Argument." 1996. bit.ly/1G9Yhk7.

———. "Pascal's Mugging." *Analysis* 69 (2009): 443–445.

———. *Superintelligence: Paths, Dangers, Strategies.* New York: Oxford University Press, 2014.

Bradley, Darren. "Bayesianism and Self-Locating Beliefs or Tom Bayes Meets John Perry." PhD diss., Stanford University, July 2007.

Bradley, Darren, and Branden Fitelson. "Monty Hall, Doomsday and Confirmation." *Analysis* 66 (2003): 23–31.

Brantley, Ben. "Review: 'Hamilton,' Young Rebels Changing History and Theater." *New York Times,* August 6, 2015.

Bresiger, Gregory. "4 of 5 Musicals Failed Their Investors." *New York Post,* January 25, 2015.

Brin, Glen David. "The 'Great Silence': The Controversy Concerning Extraterrestrial Intelligent Life." *Quarterly Journal of the Royal Astronomical Society* 24 (1983): 283–309.

Brown, Charles D. "John D. Barrow and Frank J. Tipler's 'The Anthropic Cosmological Principle.'" *Reason Papers* 13 (1988): 217–233.

Browne, Malcolm W. "Limits Seen on Human Existence." *New York Times,* June 1, 1993, C1–C7.

Brumfiel, Geoff. "Weasel Apparently Shuts Down World's Most Powerful Particle Collider." NPR, April 19, 2016. n.pr/1VXHl6l.

Carr, Michael. "Turtle Trading: A Market Legend." *Investopedia,* February 13, 2018.

Carroll, Rory. "Elon Musk's Mission to Mars." *Guardian,* July 17, 2013.

Carter, Brandon. "The Anthropic Principle and Its Implications for Biological Evolution." *Philosophical Transactions of the Royal Society of London Series A* 310, no. 1512 (1983): 347–363.

———. "Anthropic Principle in Cosmology." June 2004. arXiv:gr-qc/0606117v1.

———. "Large Number Coincidences and the Anthropic Principle in Cosmology." In M. S. Longair, ed., *Confrontation of Cosmological Theories with Observational Data, Proceedings of the Symposium,* Kraków, September 10–12, 1973, IAU Symposium No. 63, Dordrecht: D. Reidel, 1974, 291–298.

Caves, Carleton M. "Predicting Future Duration from Present Age: A Critical Assessment." *Contemporary Physics* 41 (2000): 143.

———. "Predicting Future Duration from Present Age: Revisiting a Critical Assessment of Gott's Rule." 2008. bit.ly/2sjdzPi.

Cellan-Jones, Rory. "Stephen Hawking Warns Artificial Intelligence Could End Mankind." BBC News, December 2, 2014. bbc.in/1vgH80r.

Chown, Marcus. "Dying to Know: Would You Lay Your Life on the Line for a Theory?" *New Scientist,* December 20/27, 1997, 50–51.

Clark, Nicola, and Dennis Overbye. "Scientist Suspected of Terrorist Ties." *New York Times,* October 9, 2009.

Clifford, Catherine. "Mark Zuckerberg Doubles Down Defending A.I. After Elon Musk Says His Understanding of It Is 'Limited.'" CNBC, July 26, 2017.

Coleman, Sidney, and Frank De Luccia. "Gravitational Effects on and of Vacuum Decay." *Physical Review D* 21 (1980): 3305–3315.

Collins, Daniel P. "William Eckhardt: The Man Who Launched 1,000 Systems." *Futures,* March 1, 2011.

Connors, Richard J. *Warren Buffett on Business: Principles from the Sage of Omaha.* Hoboken, NJ: Wiley, 2010.

Coughlan, Maggie. "Brad Pitt Spent Thanksgiving in London on Set." *People,* November 26, 2012.

Craig, Andrew. "Astronomers Count the Stars." BBC News, July 22, 2003.

Deutsch, David. *The Beginning of Infinity.* New York: Allen Lane, 2011.

Dicke, R. H. "Dirac's Cosmology and Mach's Principle." *Nature* 192 (1961): 440–441.

Dieks, Dennis. "Doomsday—Or: The Dangers of Statistics." *Philosophical Quarterly* 42 (1992): 78–84.

———. "The Probability of Doom." April 24, 2001. bit.ly/2joQBCm.

———. "Reasoning About the Future: Doom and Beauty." *Synthese* 156 (2007): 427–439.

DNews, uncredited. "The Soviet Particle Accelerator That Time Forgot." *Seeker.* February 24, 2011. bit.ly/2JQX2ZQ.

Dodd, Matthew S., Dominic Papineau, Tor Greene, et al. "Evidence for Early Life in Earth's Oldest Hydrothermal Vent Precipitates." *Nature* 543 (2017): 60–64.

Dowd, Maureen. "Elon Musk's Billion-Dollar Crusade to Stop the Apocalypse." *Vanity Fair,* March 26, 2017.

Dyson, Freeman J. "Reality Bites" (review of John Leslie's *The End of the World*). *Nature* 380 (1996): 296.

———. "Time Without End: Physics and Biology in an Open Universe." *Reviews of Modern Physics* 51 (1979): 447–460.

Dyson, Lisa, Matthew Kleban, and Leonard Susskind. "Disturbing Implications of a Cosmological Constant." 2002. arXiv:hep-th/0208013.

Eckhardt, William. *Paradoxes in Probability Theory.* Dordrecht: Springer, 2013.

———. "Probability Theory and the Doomsday Argument." *Mind* 102 (1993): 483–488.

———. "A Shooting-Room View of Doomsday." *Journal of Philosophy* 94 (1997): 244–259.

Eddington, Arthur. *The Nature of the Physical World.* Cambridge: Cambridge University Press, 1928.

———. *The Philosophy of Physical Science.* Cambridge: Cambridge University Press, 1939.

Einstein, Albert. *Sidelights on Relativity.* London: Methuen, 1922.

Elga, Adam. "Defeating Dr. Evil with Self-Locating Belief." *Philosophy and Phenomenological Research* 69 (2004): 383–396.

———. "Self-Locating Belief and the Sleeping Beauty Problem." *Analysis* 60 (2000): 143–147.

Evening Standard, uncredited. "The World Will End in 2060, According to Newton." June 18, 2007.

Faith, Curtis. "The Original Turtle Trading Rules." 2003. bit.ly/1GHIrIw.

Farley, Tim. "A Skeptical Maxim (May) Turn 75 This Week." *Skeptic,* November 4, 2014. bit.ly/2w9leot.

Feinstein, Alvan R. "Clinical Biostatistics XXXIX. The Haze of Bayes, the Aerial Palaces of Decision Analysis, and the Computerized Ouija Board." *Clinical Pharmacology and Therapeutics* 21 (1977): 482–496.

Fermi, Enrico. "The Future of Nuclear Physics." In J. W. Cronin, ed., *Fermi Remembered.* Chicago: University of Chicago Press, 2004.

Ferris, Timothy. "How to Predict Everything." *The New Yorker,* July 12, 1999, 35.

Feynman, Richard P. *The Meaning of It All: Thoughts of a Citizen Scientist.* Reading, MA: Addison-Wesley, 1998.

———. *QED: The Strange Theory of Light and Matter.* Princeton: Princeton University Press, 1985.

Feynman, Richard P., as told to Ralph Leighton. *Surely You're Joking, Mr. Feynman! Adventures of a Curious Character.* New York: W. W. Norton, 1985.

Fingas, Jon. "Putin Says the Country That Perfects AI Will Be 'Ruler of the World.'" *Engadget,* September 4, 2017. engt.co/2HPsATL.

Flam, F. D. "The Odds, Continually Updated." *New York Times,* September 29, 2014.

Franceschi, Paul. "A Third Route to the Doomsday Argument." *Journal of Philosophical Research* 34 (2009): 263–278.

Friedman, Milton, and Rose Friedman. *Two Lucky People: Memoirs.* Chicago: University of Chicago Press, 1998.

Friend, Tad. "Sam Altman's Manifest Destiny." *The New Yorker,* October 10, 2016.

Gardner, Martin. *Fads and Fallacies in the Name of Science.* New York: Putnam, 1952.

———. "WAP, SAP, FAP & PAP." *New York Review of Books* 33 (1986): 22–25.

Gerig, Austin. "The Doomsday Argument in Many Worlds." September 27, 2012. arXiv:1209.6251v1 [physics.pop-ph].

Glanz, James. "Point, Counterpoint and the Duration of Everything." *New York Times,* February 8, 2000.

Goldman, Albert. "Lindy's Law." *New Republic,* June 13, 1964, 34–35.

Good, Irving John. "Speculations Concerning the First Ultraintelligent Machine." *Advances in Computers* 6 (1965): 31–88.

Goodman, Steven N. "Future Prospects Discussed." *Nature* 368 (1994): 106.

Gorroochurn, Prakash. "Errors of Probability in Historical Context." *American Statistician* 65 (2011): 246–254.

Gosse, Philip Henry. *Omphalos: An Attempt to Untie the Geological Knot.* London: John Van Voorst, 1857.

Gott, J. Richard, III. "The Chances Are Good You're Random" (letter). *New York Times,* July 27, 1993 (a).

———. "Creation of Open Universes from de Sitter Space." *Nature* 295 (1982): 304–307.

———. "Future Prospects Discussed." *Nature* 368 (1994): 108.

———. "A Grim Reckoning." *New Scientist,* November 15, 1997.

———. "Implications of the Copernican Principle for Our Future Prospects." *Nature* 363 (1993): 315–319.

Grace, Caitlyn. "Anthropic Reasoning in the Great Filter." BS thesis, Australian National University, 2010. bit.ly/2jq684I.

Ha, Andrew. "Eric Schmidt Says Elon Musk Is 'Exactly Wrong' About AI." *TechCrunch,* May 25, 2018.

Hacking, Ian. "The Inverse Gambler's Fallacy: The Argument from Design. The Anthropic Principle Applied to Wheeler Universes." *Mind* 96 (1987): 331–340.

Halpern, Joseph. "The Role of the Protocol in Anthropic Reasoning." *Ergo* 2 (2015): 195–206.

Hanson, Robin. "How to Live in a Simulation." *Journal of Evolution and Technology* 7, no. 1 (2001). bit.ly/2riQThG.

———. "Must Early Life Be Easy? The Rhythm of Major Evolutionary Transitions." 1998. bit.ly/2HKKFC1.

Hawking, Stephen. *Black Holes and Baby Universes.* New York: Bantam, 1993.

Hawking, S. W., and Thomas Hertog. "A Smooth Exit from Eternal Inflation?" April 20, 2018. arXiv:1707.07702v3 [hep-th].

Hersher, Rebecca. "World's Most Destructive Stone Marten Goes on Display in the Netherlands." NPR, February 1, 2017.

Hewitt, Godfrey. "The Genetic Legacy of the Quaternary Ice Age." *Nature* 405 (2000): 907–913.

Horgan, John. "AI Visionary Eliezer Yudkowsky on the Singularity, Bayesian Brains and Closet Goblins." *Scientific American* (blog), March 1, 2016.

———. "Bayes's Theorem: What's the Big Deal?" *Scientific American* (blog), January 4, 2016.

Hume, David. *An Enquiry Concerning Human Understanding.* 1748. www.gutenberg.org/ebooks/9662.

Hunter, Matt. "Here's How One of Google's Top Scientists Thinks People Should Prepare for Machine Learning." CNBC, April 29, 2017. cnb.cx/2HN0g4l.

Hut, Piet, and Martin J. Rees. "How Stable Is Our Vacuum?" *Nature* 302 (1983): 508–509.

Iklé, Fred Charles, G. J. Aronson, and Albert Madansky. "On the Risk of an Accidental or Unauthorized Nuclear Detonation." RM-2251. Santa Monica: RAND Corporation, 1958.

Jaynes, E. T. "Prior Probabilities." *IEEE Transactions on Systems Science and Cybernetics, SSC-4* (1968): 227–241.

Jenkins, Alejandro, and Gilad Perez. "Looking for Life in the Multiverse." *Scientific American,* January 2010.

Jones, Eric M. "'Where Is Everybody?': An Account of Fermi's Question." Los Alamos, NM: Los Alamos National Laboratory, 1985.

Joy, Bill. "Why the Future Doesn't Need Us." *Wired,* April 2000.

Kahneman, Daniel, and Amos Tversky. "On Prediction and Judgment." *Oregon Research Institute Bulletin* 12, no. 4 (1972).

Keynes, John Maynard. *A Treatise on Probability.* London: Macmillan, 1921.

Khatchadourian, Raffi. "The Doomsday Invention." *The New Yorker,* November 23, 2015.

Kierland, Brian, and Bradley Monton. "How to Predict Future Duration from Present Age." *Philosophical Quarterly* 56 (2006): 16–38.

Klepper, David. "Man Recalls Day a Nuclear Bomb Fell on His Yard." *Sun News* (Myrtle Beach, SC), November 24, 2003.

Knapp, Alex. "How Much Does It Cost to Find a Higgs Boson?" *Forbes,* July 5, 2012.

Knobe, Joshua, Ken D. Olum, and Alexander Vilenkin. "Philosophical Implications of Inflationary Cosmology." *British Journal for the Philosophy of Science* 57 (2006): 47–67.

Kopf, Tomás, Pavel Krtous, and Don M. Page. "Too Soon for Doom Gloom." 1994. "Slightly revised 2012 December 21, Mayan Long Count Calendar 13.0.0.0.0, or day 1,872,000, end of the 13th b'ak'tun." arXiv:gr-gc/9407002v2.

Kosoff, Maya. "The One Technology That Terrifies Elon Musk." *Vanity Fair,* June 2, 2016.

Kragh, Helge. "Magic Number: A Partial History of the Fine-Structure Constant." *Archive for History of Exact Sciences* 57 (2003): 395–431.

Kriss, Sam. "Tech Billionaires Want to Destroy the Universe." *Atlantic,* October 13, 2016.

Lawton, John H., and Robert McCredie. *Extinction Rates.* Oxford: Oxford University Press, 1995.

Lederman, Leon. *The God Particle: If the Universe Is the Answer, What Is the Question?* Boston: Houghton Mifflin, 1993.

Lee, Jennifer Lauren. "TeV Revs Up, Operators Troubleshoot, Fight Raccoons." *Fermilab Today,* June 19, 2006. bit.ly/1QFp7yw.

Lee, Stephanie M. "This Guy Says He's the First Person to Attempt Editing His DNA with CRISPR." *BuzzFeed News,* October 14, 2017.

Lerner, Eric J. "Horoscopes for Humanity." *New York Times,* July 14, 1993.

Leslie, John. "The end of the World is not nigh" (letter). *Nature* 387 (1997): 338–339.

———. *The End of the World: The Ethics and Science of Human Extinction.* London: Routledge, 1996.

———. *Infinite Minds: A Philosophical Cosmology.* Oxford: Clarendon, 2001.

———. "Is the End of the World Nigh?" *Philosophical Quarterly* 40 (1990): 65–72.

———. "No Inverse Gambler's Fallacy in Cosmology." *Mind* 97 (1988): 269–272.

———. "Risking the World's End." *Bulletin of the Canadian Nuclear Society* 10 (1989): 10–15.

———. "The Risk That Humans Will Soon Be Extinct." *Philosophy* 85 (2010): 447–463.

———. "Time and the Anthropic Principle." *Mind* 101 (1992): 521–540.

———. *Universes.* London: Routledge, 1989 (a).

Leslie, John, and Robert Lawrence Kuhn, eds. *The Mystery of Existence: Why Is There Anything At All?* Malden, MA: Wiley-Blackwell, 2013.

Lichtenberg, Georg Christoph, trans. by R. J. Hollingdale. *The Waste Books.* New York: New York Review Books, 1990.

Liger-Belair, Gérard. *Uncorked: The Science of Champagne.* Princeton: Princeton University Press, 2004.

Linde, Andrei D. "A New Inflationary Universe Scenario: A Possible Solution of the Horizon, Flatness, Homogeneity, Isotropy and Primordial Monopole Problems." *Physics Letters B* 108 (1982): 389–393.

Mackay, Alan L. "Future Prospects Discussed." *Nature* 368 (1994): 107.

Mallah, Jacques. "Many-Worlds Interpretations Can Not Imply 'Quantum Immortality.'" 2009. arxiv.org/pdf/0902.0187.pdf.

Marchal, Bruno. "Informatique théorique et philosophie de l'esprit." In *Actes du 3ème colloque international Cognition et Connaissance.* Toulouse, 1988, 193–227.

Mata, José, and Pedro Portugal. "Patterns of Entry, Post-Entry Growth and Survival: A Comparison Between Domestic and Foreign-Owned Firms." *Small Business Economics* 22 (2004): 283–298.

McGrayne, Sharon Bertsch. *The Theory That Would Not Die: How Bayes' Rule Cracked the Enigma Code, Hunted Down Russian Submarines, & Emerged Triumphant from Two Centuries of Controversy.* New Haven: Yale University Press, 2011.

Merali, Zeeya. "Quantum Physics: What Is Really Real?" *Nature* 521 (2015): 278–280.

Metz, Cade. "Mark Zuckerberg, Elon Musk and the Feud over Killer Robots." *New York Times,* June 9, 2018.

Military History Now, uncredited. "The Men Who Saved the World—Meet Two Different Russians Who Prevented WW3." 2013. bit.ly/1PbhTli.

Moravec, Hans. *Mind Children: The Future of Robot and Human Intelligence.* Cambridge: Harvard University Press, 1988.

Mosher, Dave. "A Thermonuclear Bomb Slammed into a North Carolina Farm in 1961—and Part of It Is Still Missing." *Business Insider,* May 7, 2017.

Moskowitz, Clara. "Are We Living in a Computer Simulation?" *Scientific American,* April 7, 2016.

Neal, Radford M. "Puzzles of Anthropic Reasoning Resolved Using Full Non-Indexical Conditioning." 2006. arXiv:math/0608592v1.

Nielsen, Holger Bech. "Random Dynamics and Relations Between the Number of Fermion Generations and the Fine Structure Constants." *Acta Physica Polonica* B20 (1989): 427–468.

Nielsen, H. B., and Masao Ninomiya. "Card Game Restriction in LHC Can Only Be Successful!" October 23, 2009. arXiv:0910.0359 [physics .gen-ph].

Norton, John D. "Cosmic Confusions: Not Supporting Versus Supporting Not." *Philosophy of Science* 77 (2010): 501–523.

O'Connell, Cathal. "Can We Test for Parallel Worlds?" *Cosmos,* November 3, 2014.

Oliver, Bernard M., and John Billingham. *Project Cyclops: A Design Study of a System for Detecting Extraterrestrial Intelligent Life*. Moffett Field, CA: Stanford/NASA/Ames Research Center, 1971.

Olum, Ken D. "The Doomsday Argument and the Number of Possible Observers." October 13, 2000. arXiv:gr-qc/0009081v2.

Olum, Ken D., and Delia Schwartz-Perlov. "Anthropic Prediction in a Large Toy Landscape." May 17, 2007. arXiv:0705.2562v2 [hep-th].

Overbye, Dennis. "The Collider, the Particle and a Theory About Fate." *New York Times,* October 12, 2009.

Page, Don N. "Can Quantum Cosmology Give Observational Consequences of Many-Worlds Quantum Theory?" In C. P. Burgess and R. C. Myers, eds., *Eighth Canadian Conference on General Relativity and Relativistic Astrophysics,* Montreal. Melville, NY: American Institute of Physics, 1999, 225–232.

Page, Lewis. "Large Hadron Collider Scuttled by Birdy Baguette-Bomber." *Register,* November 5, 2009.

Papineau, David. "Why You Don't Want to Get in the Box with Schrödinger's Cat." *Analysis* 63 (2003): 51–58.

Physics World, uncredited. "Life, Longevity, and a $6000 Bet." February 11, 2000.

Piccione, Michele, and Ariel Rubinstein. "On the Interpretation of Decision Problems with Imperfect Recall." *Games and Economic Behavior* 20 (1997): 3–25.

Poundstone, William. *Are You Smart Enough to Work at Google?* New York: Little, Brown, 2012.

———. *Carl Sagan: A Life in the Cosmos*. New York: Holt, 1999.

———. *Labyrinths of Reason: Paradox, Puzzles, and the Frailty of Knowledge*. New York: Doubleday, 1988.

———. *Rock Breaks Scissors: A Practical Guide to Outguessing and Outwitting Almost Everybody*. New York: Little, Brown, 2014.

Purcell, Edward. "Radio Astronomy and Communication Through Space." In A. G. W. Cameron, ed., *Interstellar Communication*. New York: W. A. Benjamin, 1963.

Putnam, Hilary. "The Place of Facts in a World of Values." In D. Huff and O. Prewett, eds., *The Nature of the Physical Universe*. New York: Wiley, 1979, 113–140.

Rees, Martin. *Just Six Numbers: The Deep Forces That Shape the Universe*. London: Weidenfeld & Nicolson, 1999.

———. *Our Final Century: Will the Human Race Survive the Twenty-First Century?* London: Heinemann, 2003.

Reynolds, Ben. "The Lindy Effect: Triumph of the Tried & True." August 27, 2016. bit.ly/2CzwoEC.

———. "107 Profound Warren Buffett Quotes: Learn to Build Wealth." n.d. bit.ly/2riDpmR.

Rubinow, Isaac M. "Scientific Methods of Computing Compensation Rates." *Proceedings of the Casualty Actuarial Society* (1914–15): 10–23.

Sandberg, Anders, Eric Drexler, and Toby Ord. "Dissolving the Fermi Paradox." June 6, 2018. arXiv:1806.02404v1 [physics.pop-ph].

Saunders, Tristram Fane. "10 Things You Didn't Know About The Mousetrap." *Telegraph*, November 25, 2015.

Scharf, Caleb. *The Copernicus Complex*. New York: Scientific American/Farrar, Straus and Giroux, 2014.

Searle, John. *Mind, Language, and Society*. New York: Basic Books, 1999.

Sebens, Charles T., and Sean M. Carroll. "Self-Locating Uncertainty and the Origin of Probability in Quantum Mechanics." *British Journal for the Philosophy of Science* 69 (2015): 25–74.

Selby, Andrew. "Market Wizard William Eckhardt." *Don't Talk About Your Stocks* (blog and podcast), January 22, 2013. bit.ly/1AgwK9G.

Shikhovtsev, Eugene. "Biographical Sketch of Hugh Everett, III." 2003. bit.ly/2HRqnmz.

Sober, Elliott. "An Empirical Critique of Two Versions of the Doomsday Argument—Gott's Line and Leslie's Wedge." *Synthese* 135 (2003): 415–430.

Sowers, George F., Jr. "The Demise of the Doomsday Argument." *Mind* 111 (2002): 37–45.

Spiegel, David S., and Edwin L. Turner. "Bayesian Analysis of the Astrobiological Implications of Life's Early Emergence on Earth." *PNAS* 109 (2012): 395–400.

Stevenson, Seth. "A Rare Joint Interview with Microsoft CEO Satya Nadell and Bill Gates." *Wall Street Journal,* September 25, 2017.

Stigler, Stephen M. "The True Title of Bayes's Essay." *Statistical Science* 28 (2013): 283–288.

———. "Who Discovered Bayes's Theorem?" *American Statistician* 37 (1983): 290–296.

Taleb, Nassim Nicholas. *Antifragile: Things That Gain from Disorder.* New York: Random House, 2012 (a).

———. "The Surprising Truth: Technology Is Aging in Reverse." *Wired,* December 21, 2012.

Tegmark, Max. "The Interpretation of Quantum Mechanics: Many Worlds or Many Words?" 1997. bit.ly/2rjh1ZS.

———. *Our Mathematical Universe: My Quest for the Ultimate Nature of Reality.* New York: Knopf, 2014.

Tegmark, Max, and Nick Bostrom. "Is a Doomsday Catastrophe Likely?" *Nature* 438 (2005): 754. An extended version (2005a) is at bit.ly/2smf3Z1.

Thompson, Clive. "If You Liked This, You're Sure to Love That." *New York Times,* November 21, 2008.

Thwaite, Ann. *Glimpses of the Wonderful: The Life of Philip Henry Gosse, 1810–1888.* London: Faber & Faber, 2002.

Tuttle, Steve. "A Close Call." East Carolina University News Service, March 27, 2013. bit.ly/2rkv0za.

Tyson, Neil deGrasse, Michael A. Strauss, and J. Richard Gott. *Welcome to the Universe: An Astrophysical Tour.* Princeton: Princeton University Press, 2016.

Ulam, Stanislaw. "Tribute to John von Neumann." *Bulletin of the American Mathematical Society* 5 (1958): 1–49.

van der Vat, Dan. "Jack Good" (obituary). *Guardian,* April 28, 2009.

Varandani, Suman. "Stephen Hawking Puts an Expiry Date on Humanity." *International Business Times,* November 16, 2016.

Von Foerster, Heinz, Patricia M. Mora, and Lawrence W. Amiot. "Doomsday: Friday, November 13, AD 2026." *Science* 132 (1960): 1291–1295.

Wade, Nicholas. "Genome Study Provided a Census of Early Humans." *New York Times*, January 18, 2010.

Wearing, J. P. *The London Stage: A Calendar of Plays and Players.* Metuchen, NJ: Scarecrow Press, 1976–1993. Published in seven volumes, each covering a decade of London productions.

Weinberg, Steven. *Dreams of a Final Theory.* New York: Pantheon, 1992.

Wells, Willard. *Apocalypse When: Calculating How Long the Human Race Will Survive.* Chichester, UK: Praxis, 2009.

Whitrow, G. J. "Why Physical Space Has Three Dimensions." *British Journal for the Philosophy of Science* 6 (1955): 13–31.

Wigner, Eugene P., and A. Szanton. *The Recollections of Eugene P. Wigner.* New York: Plenum, 1992.

Williams, John Burr. *The Theory of Investment Value.* Cambridge: Harvard University Press, 1938.

Yudkowsky, Eliezer. "Artificial Intelligence as a Positive and Negative Factor in Global Risk." In Nick Bostrom and Milan M. Cirkovic, eds., *Global Catastrophic Risks.* New York: Oxford University Press, 2008, 308–345.

Zuboff, Arnold. "One Self: The Logic of Experience." *Inquiry* 33 (1990): 39–68.

———. "Time, Self and Sleeping Beauty." 2008. bit.ly/2HLZzrQ.

Zullo, Robert. "The Future of History: New Direction for Colonial Williamsburg Met with Praise, Backlash." *Richmond Times-Dispatch*, March 12, 2016.

Index

Note: Italic page numbers refer to illustrations.

Index

Index

Index

Index

About the Author

WILLIAM POUNDSTONE is the author of sixteen previous books, including *Are You Smart Enough to Work at Google?*, *Fortune's Formula,* and *How Would You Move Mount Fuji?*. He has written for the *New York Times, Harper's, Harvard Business Review,* and the *Village Voice,* among other publications, and is a frequent guest on TV and radio. He lives in Los Angeles. Follow Poundstone on Twitter (@WPoundstone) and learn more at his website, william-poundstone.com.

Also by William Poundstone

Head in the Cloud

Why Knowing Things Still Matters When Facts Are So Easy to Look Up

"True fact: People who read Poundstone's extraordinary books are smarter (and happier) than people who don't."

—Seth Godin, author of *The Icarus Deception*

Rock Breaks Scissors

A Practical Guide to Outguessing and Outwitting Almost Everybody

"A smart, engagingly written account of how to capitalize on other people's predictability...An enlightening book."

—David Pitt, *Booklist*

Are You Smart Enough to Work at Google?

Trick Questions, Zen-like Riddles, Insanely Difficult Puzzles, and Other Devious Interviewing Techniques You Need to Know to Get a Job Anywhere in the New Economy

"An enjoyably brain-stretching account of the world's toughest, most mischievous job-interview questions." —David Rowan, *Wired*

How Would You Move Mount Fuji?

Microsoft's Cult of the Puzzle: How the World's Smartest Companies Select the Most Creative Thinkers

"A charming Trojan horse of a book. Revealing the tricks to Microsoft's notorious hiring challenges...Poundstone shows how puzzles can—and cannot—identify the potential stars of a competitive company."

—Tom Ehrenfeld, *Boston Globe*